"This innovative volume brings together a compelling, expansive and macro-level concept – imagining destination futures – along with rich empirical evidence from three contrasting Asia Pacific settings. The future orientation is invaluable for students and scholars who are challenging long-established certainties with thoughtful guiding principles to guide the turbulence ahead facing tourism."

Brian King, *Professor and Head, Department of Hospitality, Hotel Management and Tourism, Texas A&M University, USA*

"Tourism is confronting numerous challenges that have increased the vulnerability of destinations worldwide, and it is crucial to rethink about the future of tourism destinations. This book explores *possible*, *probable*, and *preferable futures* by examining three renowned tourism destinations and proposing viable solutions. I highly recommend this book as an invaluable resource for both students and researchers in the field."

Professor S. Mostafa Rasoolimanesh, *Professor of Sustainable Tourism, School of Business and Law, Edith Cowan University, Australia*

The Possible, Probable and Preferable Futures of Tourism Destination Management

Topical and novel, this book critically explores the possible, probable and preferable futures of tourism destinations, exposing the reader to numerous issues and challenges related to current, global societies and geopolitical settings and adaptive strategies to mitigate such challenges at all levels and contexts.

This is based on unique longitudinal research over 15 years comparing and contrasting destinations in mature urban metropolitan destinations, (emerging) island destinations with many developments and fast developing destinations in the Asia Pacific (Melbourne, Penang and Chiang Mai). The interconnected facets of a community, internal issues and challenges, external drivers of change, and government interventions in envisioning the future of these destinations are explored along with core issues of climate change, sustainability, ecological and socio-ecological dimensions of destinations. This significant volume offers insight into the environmental, political and economic challenges facing destinations and possible ways of alleviating these at different scales and contexts, informing future policy and development.

This book is essential reading for all students, researchers and academics interested in tourism futures.

Rami K. Isaac is currently an associate professor in the Faculty of Tourism and Hospitality at Bethlehem University, Palestine, and associate professor at Breda University of Applied Sciences.

Erdinç Çakmak is a professor at Breda University of Applied Sciences (formerly NHTV) and has developed, coordinated and taught a range of courses on destination management, context and field research at introductory and advanced levels.

Richard Butler is Emeritus Professor in the Business School at the University of Strathclyde.

Contemporary Geographies of Leisure, Tourism and Mobility
Series Editor: C. Michael Hall, *Dean's Chair in Marketing, Sustainability and Society, Massey Business School, Te Kura Whai Pakihi, Massey University, Albany, New Zealand*

The aim of this series is to explore and communicate the intersections and relationships between leisure, tourism and human mobility within the social sciences.

It will incorporate both traditional and new perspectives on leisure and tourism from contemporary geography, e.g. notions of identity, representation and culture, while also providing for perspectives from cognate areas such as anthropology, cultural studies, gastronomy and food studies, marketing, policy studies and political economy, regional and urban planning, and sociology, within the development of an integrated field of leisure and tourism studies.

Also, increasingly, tourism and leisure are regarded as steps in a continuum of human mobility. Inclusion of mobility in the series offers the prospect to examine the relationship between tourism and migration, the sojourner, educational travel, and second home and retirement travel phenomena.

The series comprises two strands:

Contemporary Geographies of Leisure, Tourism and Mobility aims to address the needs of students and academics, and the titles will be published simultaneously in hardback and paperback. Titles include:

Tourism and Animal Ethics 2nd Edition
David A. Fennell

Routledge Studies in Contemporary Geographies of Leisure, Tourism and Mobility is a forum for innovative new research intended for research students and academics, and the titles will initially be available in hardback only.
Titles include:

The Possible, Probable and Preferable Futures of Tourism Destination Management
Rami K. Isaac, Erdinç Çakmak and Richard Butler

For more information about this series, please visit: www.routledge.com/Contemporary-Geographies-of-Leisure-Tourism-and-Mobility/book-series/SE0522

The Possible, Probable and Preferable Futures of Tourism Destination Management

Rami K. Isaac, Erdinç Çakmak and
Richard Butler

Routledge
Taylor & Francis Group
LONDON AND NEW YORK

First published 2026
by Routledge
4 Park Square, Milton Park, Abingdon, Oxon OX14 4RN

and by Routledge
605 Third Avenue, New York, NY 10158

Routledge is an imprint of the Taylor & Francis Group, an informa business

ISBN: 9781032943428 (hbk)
ISBN: 9781032943442 (pbk)
ISBN: 9781003570240 (ebk)

DOI: 10.4324/9781003570240

Typeset in Times New Roman
by Newgen Publishing UK

Contents

Figures

Table

About the Authors

The co-authors of this volume have a great deal of experience in the field of tourism development and management. The first two co-authors are part of the teaching staff and supervisors of the field research.

Rami K. Isaac has more than 20 years of experience in tourism studies. Born in Palestine, Rami Isaac completed his undergraduate studies in The Netherlands, graduate studies in the United Kingdom and earned his Ph.D. from the University of Groningen, in spatial sciences, in The Netherlands. He is currently an associate professor in the Faculty of Tourism and Hospitality at Bethlehem University, Palestine, and at Breda University of Applied Sciences. He was the former president of the Research Committee 50 (RC50) on International Tourism, International Sociologist Association (ISA) (2014–2018) and currently he is the Vice-President of RC50 Tourism (2018–2027). His research interests are in the area of tourism development and management, cultural heritage/dark tourism and political aspects of tourism. He has co-authored more than 80 academic papers, and book chapters on tourism and political (in)stability, occupation, tourism and war, violence and transformational tourism. He is also the co-editor of the books *The Politics and Power of Tourism in Palestine* (2016); *Tourism and Hospitality in Conflict-ridden Destinations* (2019); *Tourism Interventions: Making or Breaking Places* (2024); *Tourism, Safety, Security and Resilience: A Community Integrated Approach* (2024) by Routledge and co-editor of the *Changing Practices of Tourism Stakeholders in COVID-19-Affected Destinations* (2023) by Channel View.

Erdinç Çakmak, born in Turkey, obtained his Ph.D. in Social Sciences from Wageningen University and Research. Since 2007 he has been working at Breda University of Applied Sciences (formerly NHTV) and has developed, coordinated and taught a range of courses on destination management, context and field research at introductory and advanced levels. He has co-authored more than 30 academic papers and book chapters, co-edited special volumes and books and apart from that co-chaired international conferences. He has been the vice president of the international tourism group RC50 at the ISA since 2014. He is the co-editor of the recently (2021) published volume *Tourism Paradoxes: Contradictions, Controversies and Challenges* by Channel View Publications.

Richard Butler, born in the United Kingdom, completed degrees in geography at Nottingham University (B.A. Hons.) and Glasgow University (Ph.D.) before moving to Canada in 1967 where he was at the University of Western Ontario (London, Canada) for 30 years. There he taught tourism leisure and recreation geography, completing his term as Chair of the Department of Geography. He returned to the United Kingdom in 1997 as Deputy Head (Research) at the University of Surrey and then moved to the University of Strathclyde in 2005, where he is an Emeritus Professor in the Business School. He has published 25 books on tourism, is a past president of the International Academy for the Study of Tourism and the Canadian Leisure Studies Association and on the editorial boards of several tourism journals. He has published many articles and chapters on tourism development, island tourism, tourism and war, religion and resilience, and has been a visiting professor at Breda University on several occasions. In 2016, he was named the UNWTO Ulysses Laureate for his contributions to tourism knowledge.

Preface

This book aims to furnish readers with vital practical insights into the domains of tourism and future studies. The chapters are structured to examine these topics through future studies, based on the *possible, probable* and *preferable futures of tourism destinations*, which entails a comprehensive consideration of the interconnected facets of a community, internal issues and challenges, external drivers of change and government interventions in envisioning the future of these destinations. The book is conceptualised around several thematic areas.

Nowadays we are experiencing a period of intense and complex relations involving tourism issues and challenges because of factors influencing tourism at the national and international levels. By exploring the futures of tourism in the contexts of possible, probable and preferable futures, we offer current and future politicians, managers, academics, practitioners and students insights into key international tourism case studies, issues, trends, forces and actors through these international tourism destinations. This volume will also provide Destination management organisations (DMOs), tourism boards and local tourism authorities with information for developing tourism policies and plans in order to increase their competitiveness as tourism destinations.

We view this level of presentation of tourism futures as apposite for several reasons. One of these reasons is related to a fieldwork assignment for the postgraduate students in the MA in Tourism Destination Management (TDM) programme at Breda University of Applied Sciences, whereby they spend three months in Asia Pacific. The assignments were created by lecturers from Breda University of Applied Sciences MA in TDM programme in collaboration with stakeholders involved in tourism development at three different locations. These destinations were chosen for a variety of reasons, including TDM staff's expertise in the different places, relationships with businesses, governmental organisations and academics in the areas, and familiarity with the various stages of tourism development in both markets and economies. The three destinations picked differ in the following ways: (1) a mature urban metropolitan destination, Melbourne,

Australia; (2) an (emerging) island destination with many developments, Penang, Malaysia; (3) a fast developing inland destination, Chiang Mai, Thailand. The purpose is for students to become engaged in and familiar with the various and complicated contextual realities of destination management in 'urban,' 'island' and 'rapidly' developing tourism locations.

We have organised our approach to the broad subject of tourism futures in nine chapters, each with a specific case study report. Chapter 1 provides an introduction and an overview of the book, its aims and key current issues affecting tourism and the future of tourism. Chapters 2, 3 and 4 deal with the possible futures: Collect and assess: Match and possible futures. Chapters 5, 6 and 7 focus on the probable future: Analyse and evaluate: Match and probable future. Chapters 8, 9 and 10 address the preferable future: Consult and advise: Match and preferable future. The final chapter (Chapter 11) identifies common themes and issues gathered from the preceding chapters and produces conclusions concerning the nature of and implications to be drawn from the adjustment of practices of future scenarios.

Acknowledgements

The authors would like to thank the staff at Routledge for their patience and support throughout the preparation of the book proposal and submission of the manuscript, and in particular Emma Travis and Harriet Cunningham for their continued encouragement and support. Furthermore, we would like to thank the TDM students who have contributed to this book over the last two years through their analytical work.

1 Introduction

The aim of this book is to examine the possible, probable and preferable futures of tourism destinations. Recent environmental, political and economic crises have revealed the vulnerability of tourism across different regions and environments around the globe. Tourism and the related wider socio-ecological systems are currently facing important environmental challenges at different scales and contexts. The introduction of concepts like Anthropocene and Capitolocene in current tourism research reflects a shift in current conceptual and empirical studies in the field. The increasing socio-ecological vulnerabilities of destinations across the world indicate a need for a much-enhanced rethinking of destination futures and resilience at the regime level, including consideration of moving from the perpetual growth approach that has characterized tourism over the last century to a more sustainable one to match the changing definitions of success for tourism destinations (Dwyer, 2024). On the one hand, current steering modes of governance and decision-making need to be redesigned to address uncertainties and policy issues. On the other hand, businesses and organisations must embrace alternative approaches in tourism and resource use at the destination level to mitigate and adapt to the impacts of climate change and other environmental influences and changing social–cultural preferences. It is important to reframe ecological and socio-ecological dimensions of destination futures beyond the hegemonic depoliticised imaginary of the climate crisis to effectively bring about change for the better.

The COVID-19 global pandemic and its impacts on tourism following the lockdowns represented a potential turning point in the study of tourism policy and planning. As Higgins-Desbiolles (2021) has observed, academic debate in the field saw a polarisation between those advocating a much-needed reform towards more responsible and sustainable trajectories in tourism (e.g. Brouder, 2020) and more precautionary approaches (Hall et al., 2021). Hall and Seyfi (2021) discussed the impact of COVID-19 on tourism and the potentialities of de-growth and slowing down practices in acknowledgement of the environmental and social pressures destinations have been coping with prior to the pandemic. However, they noted that "many of the government responses to COVID-19 are not geared towards green responses but are instead meant to reinforce business as usual or worse" (Hall & Seyfi, 2021, p. 225). The latter comment echoes Amore and Adie's (2021) findings

DOI: 10.4324/9781003570240-1

across urban destinations and the prevailing rhetoric of swift recovery following the pandemic.

More rarely are regional tourism policy initiatives "a proactive decision as the result of an awareness of changes in the tourism business environment" (Hall, 2008, p. 104). According to Amore and Hall (2017), crises and neoliberal-driven development discourses in tourism culminate in a hyper-neoliberal spiral that "dominates the socio-technical system and frames the policy agenda that affect directly and indirectly impact tourism destinations and DMOs and sets destinations on particular development trajectories" (Amore & Hall 2017, 12).

Although there were pleas for alternative policies towards a just pathway for tourism in respect of communities and ecological limits (Higgins-Desbiolles, 2020, 2021), there was "no evidence-based strategy for climate change mitigation, and an overall silence regarding pandemic and other risks the global tourism system imposes on itself and the global economy" (Gössling, Scott & Hall, 2021, p. 13). The academic arguments over the ideal balance between economic development and sustainability (Higgins-Desbiolles, 2020) did not effectively enter public policy arenas and mirror the lack of progress in addressing the climate crisis in tourism policymaking and other 'wicked problems' facing tourism and tourism destinations.

In recent years, in part as a result of the COVID-19 pandemic, there has been a growing refocus on the importance of community-centred policy and planning in tourism. This builds on the work of some scholars and residents of destinations who have long been advocating for a local turn in tourism development which places the interests of the community at the centre of the decision-making process and could help to mitigate such issues as overtourism. Calls for greater equality, distributive justice and enhanced social sustainability highlight the necessity for an agency-driven rethinking and rightsizing of tourism at the local destination level, along with call for a change in focus towards quality of life of destination residents and tourists (Dwyer, 2023).

Climate change will affect every destination and community but it will do so unequally. Some will be better placed to adapt than others. What are the implications of this, and how can we take steps to ensure no destination is left vulnerable? It is surely no exaggeration to say that, over the next three decades, our generation will witness some of the most momentous changes in human history. The world will likely experience a rapid energy revolution as fossil fuels are abandoned for sustainable alternatives, along with game-changing technological breakthroughs to rival those of the past 30 years and, even if global heating is kept to within 1.5 degrees (the science-based goal of the Paris Agreement), there will still be significant shifts in climate patterns, a rise in sea levels and more extreme weather events. In tourism, if current trends continue, there will also be a doubling of the number of trips taken.

Thinking about the future of destinations, it is clear that some destinations are better placed than others to have a 'good transition.' All destinations are of course different, and as such the challenges they face and their ability to decarbonise and adapt will vary considerably. This raises questions about how best to build future scenarios without leaving any destination behind, and how to ensure a fair approach to future-proofing that doesn't further exacerbate existing inequalities. As a result, various authors have called upon destination management organisations (DMOs) to increasingly pivot from place marketing to place management as a policy priority, embracing social responsibility towards both visitors and locals, stewardship of local culture and resources and sustainability (Morgan, 2012; Morrison, 2013; Pechlaner, Zacher, Eckert & Petersik, 2019). For many DMOs, the broader responsibilities of destination stewardship and a shifting perspective from quantity to quality have significantly changed both their strategies and actions. The involvement of the local society has come to play an important role, since informed and involved residents often have a more accurate but still positive perception of tourism, leading to the enhanced well-being of a destination's residents and their quality of life (Šegota, Mihalič & Kuščer, 2017; Dwyer, 2024).

Recent environmental, political and economic crises have shed further light on the vulnerability of tourism across different regions and environments around the globe. From a public policy perspective, legislation, regulations and direct interventions of national, regional and local governments in tourism have long been used to address various social, economic, environmental and technological issues impacting businesses and individuals (Hall & Jenkins, 1995). However, evidence shows that many of these interventions in both developed and developing countries reflect the proliferation of neoliberal and hyper-neoliberal ideologies emphasising economic growth and laissez faire policies that reinforce the role of 'the market' as a governance mechanism (Duffy, 2015; Torkington, Stanford & Guiver, 2020). Public policy in tourism predominantly consists of reactive, ex-post policy actions and deliberations that are, in part, a response to the inactions of governments in the first place (Dodds & Butler, 2008).

The Glasgow Declaration, signed by over 500 tourism groups, intends to cut emissions in half by 2030 and eliminate them by 2050. To achieve the Glasgow Declaration's goals, a number of significant measures and interventions are required (Peeters & Papp, 2023), including lowering the number of flights and upgrading Europe's train network. As Postma, Hartman and Yeoman (2024, p. 2) state 'We live in a VUCA world': volatile, uncertain, complex and ambiguous (see also Hartman, 2023). This implies we must accept the reality of constant, unpredictable change and a lack of clarity about how interlinked elements and forces form and change, resulting in an environment that is difficult to forecast, interpret and analyse (Major & Clarke, 2022).

With rising globalisation, the digitalisation of information, the international nature of tourism and the extraordinarily fast rate of change in society, it is critical for tourism destinations to plan and prepare appropriately and responsibly for the future. The two most popular techniques are to plan in a voluntaristic manner, claiming that we may pick our future and plan it rationally, or, to adopt a deterministic manner, assuming that the future is determined by certain forces and events, which is usual in prognosis planning (Enger, Sandvik & Iversen, 2015).

The research discipline known as future studies has only emerged in the last four to five decades. To oppose the assumption of 'just one future,' futures studies is a plural phrase rather than the singular 'future' studies, the latter having both conceptual constraints and political implications. This futures pluralisation opens up the possibility of imagining and producing 'alternative' and even 'preferred' futures. While it is commonly assumed that future studies are attempts to 'predict the future' based on extrapolation from current trends, and global drivers of change, a review of future studies literature reveals that there has been a shift from what Inayatullah (2013) refers to as single-point forecasting – often as a precise prediction – to scenario planning, which embraces not only one outcome but several. Tourism, being highly dynamic, is a process and as such it is hard to predict the future.

Forecasting and prediction are now regarded as naive scientific pursuits by futures researchers, and according to Strand (1999) the purpose of futures studies is to make more informed decisions and choices while attempting to manage change processes rather than to be a prediction engine. Bergman, Karlsson and Axelsson (2010) provide a valuable typology of future studies that incorporates four paradigms of thought: prediction, prognosis, utopia and science fiction. Predictions typically represent the goals of a scientific process and can thus be associated with a certain method or procedure. Prognoses differ from explanations in that they do not make explanatory claims regarding future occurrences and situations. They are associated with a belief that 'forecasts' are plausible. A common form of prognosis is to extrapolate using empirical, usually statistical, trends. The statistical methods can be very sophisticated, but the underlying assumption is that the future will continue to follow the direction already established.

Future studies create alternative futures by making basic assumptions problematic. Through questioning the future, emerging issues analysis and scenarios, the intention is to move out of the present and create the possibility for new futures (Inayatullah, 2013).

Scenario Planning

Scenario planning is a popular approach for addressing uncertainty in strategic decision-making (Postma et al., 2024, p. 10). Scenario planning, which began as an open and uncertain attempt, has evolved into various schools and is now widely employed in research and practical situations (López-Rodríguez et al., 2023). The use of scenario planning is addressed in the literature under different names: scenario development, scenario planning and scenario thinking (Oner, 2010; Varum & Melo, 2010) or referring to a particular approach to devise scenarios, such as participatory scenario planning (Poskitt, Waylen & Ainslie, 2021).

Based on both qualitative and quantitative research methodologies, a wide range of scenario strategies have been developed. Although there appear to be few similarities, as Kosow and Gaßner (2008) highlight, there are some common characteristics of scenarios, namely: (1) scenarios are not a comprehensive image of the future, but rather hypothetical sequences of events constructed for focusing on causal processes and decision points; (2) scenarios are based on certain identified key factors and drivers of change; and (3) scenarios are based on certain identified key factors and thus do not represent the future as a whole, but rather a possible future construct of key factors. Kosow and Gaßner (2008) classified scenarios according to their time horizons as short term (up to 10 years), medium term (up to 25 years) and long term (more than 25 years).

Because of their utility in times of ambiguity and complexity, scenario-planning approaches are becoming increasingly popular in today's world of unpredictability, innovation and change. Planning scenarios encourage strategic thought and aid in overcoming thinking constraints by generating multiple futures.

Using scenarios or scenario planning is one technique to influence the objective of future studies. In the postmodern era of uncertainty and contingency, scenario thinking and planning are increasingly being used in the public and private sectors for corporate and government decision-making (Yeoman et al., 2012). Scenario planning supports the formation of a larger vision of the terrain, facilitates innovative thinking and promotes action by facilitating strategic interactions among varied stakeholders and embracing the complexity of their multiple perspectives (Yeoman et al., 2012). Scenarios are the outline of some features of the future, and a scenario often refers to an outline of the storyline of a dramatic work, the script of a motion picture or a television programme (Joseph, 2000). Khan is considered one of the founders of future studies and of scenario planning. He defines a scenario as "a set of hypothetical events set in the future constructed to clarify a possible chain of causal events as well as their decision points" (Kahn & Wiener, 1967; Nyaupan & Buzinde, 2017).

Scenarios help us make sense of these uncertainties and challenge the beliefs about what the future holds. We can all agree that some aspects of the future are predetermined – they are the unavoidable outcomes of past events. Scenarios do not represent projections, predictions or preferences. Rather, they are cohesive and plausible stories that describe several routes that lead to diverse futures. The process of creating and deploying scenarios is just as significant as the situations themselves. Excellent scenarios help connect plans with a vision, allowing society to more clearly challenge planning assumptions, benchmark goals and plan for potential macro risks. Scenarios are also a description of a future situation and the sequence of events that allows one to move from the current to the future situation (Godet, 2000). Scenarios are sometimes characterized as various futures that result from a combination of trends, drivers of change and policies (Fontela & Hingel, 1993), but they also have to take into consideration the recent environmental, political and economic crises that have shed further light on the vulnerability of tourism across different regions and environments. Tourism and the wider socio-ecological systems are currently facing important environmental challenges at different scales and contexts. Scenario planning enables society to anticipate and innovate for the future (Hiltunen, 2009). Scenarios depict the overall state of the environment and highlight the interplay between various trends, forces and events in the future (Martino, 2003).

Lindgren and Bandhold (2003) offer several explanations for the success of the scenario planning method. For starters, scenario planning provides a framework for reducing complexity to a finite number of divergent possibilities. Second, by providing team members with a collectively acknowledged and agreed structure for thinking beyond what is already known, the scenario framework allows for more efficient communication. Third, stories are easily understood by the human brain. The narrative thinking used in scenario thinking corresponds to how the brain operates, hence increasing the brain's capacity to process information. Ultimately, the brain may be educated to conceive the inconceivable by forcing it to think in fundamentally different directions. Successful scenarios must make sense to the important actors and be relevant to them. They must be plausible to stakeholders and allow them to envision themselves in the scenarios described. Simultaneously, the scenarios must challenge these stakeholders' minds; they must be novel and innovative, and they must move thinking beyond the framework of business as usual (Yeoman et al., 2012, p. 4).

The use of scenario planning has increased significantly during the last decade (Amore & Adie, 2021; Enger et al., 2015; Rigby & Bilodeau, 2007). Research indicates that there is a correlation between the adoption of scenario planning techniques and uncertainty, unpredictability and instability of the overall business environment (Malaska, Malmivirita, Meristö & Hansèn, 1984). The scenario does not predict the future, but it explores multiple plausible future situations to extend the sphere of thinking of the participants in the scenario development process (Godet, 2000; Schnaars, 1987).

A scenario method enables agencies to assess fundamentally unknown driving forces methodically and analytically, and then sketch distinct possible, probable and preferable futures of destinations. The scenario approach is a method for gaining insight into what might happen in the future in a specific subject and evaluating the change dynamics that will bring us there (Yeoman et al., 2012). Furthermore, the scenario methodology proposes a process approach that establishes a shared foundation and reality for all parties involved. The three most common scenario techniques are deductive, inductive and incremental.

The Possible, Probable and Preferable Futures

In a provocative article about the future of the healthcare system in the United States, Hancock and Bezold (1994) outlined four types of futures for healthcare organisations. One future is the 'possible future,' that is, what may happen, including events most likely to occur. The 'plausible future' is one in which trends are identified and compiled and the information is used to forecast what could happen. The 'probable future' is the future that is likely to happen based on an analysis of the present situation. Usually in this context, the future is seen as a continuation of the present, resulting in little change. The 'preferable future' is the future that one wants to happen. Vision, especially long-term vision, is needed to articulate shared goals of a destination or society. By identifying a preferable future, a proactive stance can be taken and plans for implementation of actions to achieve this future can be made. Collectively, these methods operate within the domain of three questions associated with the future (Rubin & Kaivo-Oja, 1999): possible futures – what may happen?; probable futures – what is most likely to happen?; preferable futures – what would be preferred to happen?

Using Henchey's (1978) typology of futures, there are three ways of thinking about the future:

1 **Possible future** (what may be) encompasses everything that can possibly be imagined, no matter how unlikely.
2 **Probable future** (what will likely be) is what will likely happen and is based on an examination of the current situation and an appraisal of likely trends and future developments.
3 **Preferable future** (what should be) is what is wanted to happen and falls in the realm of organisational or societal visions that 'move reality beyond the present towards the best that can be.'

The Edited Volume

This edited collection addresses a research need to understand the future of destinations. It is a book comparing and contrasting destinations in the Asia Pacific (Melbourne, Australia; Chiang Mai, Thailand; and Penang, Malaysia) based on 15 years of experience with master students' fieldwork in the Asia Pacific, during which periods of study students spend three months in Asia, one month at each destination (the first month in Melbourne, the second month in either Chiang Mai or Sri Lanka and the third month in Bali or Penang). The futures produced are based on longitudinal research over 15 years in these destinations. The results gained can be very insightful, and the implications could extend beyond geographical borders to other regions and countries. This approach is highly unique, considering the time series data and extensive history of experience in the destinations, as well as the novel incorporation of various perspectives, key drivers of change and key issues from those destinations and the researchers.

The expertise collected over the last 15 years of experience has resulted in the identification and categorisation of relevant formal and informal knowledge required to evaluate the present fit between the destination area's features/assessments/potential and the current markets. We were able to reveal the destinations' future development potential and advise on realistic, effective product-, market- and organisational initiatives. We were able to assist destinations in achieving their preferred future and improving their competitive advantage.

Structure of the Book

Part I: The Possible Futures: Collect and Asses: Match and Possible Futures

Part I includes several possible futures focusing on different areas in Melbourne, Australia, and the implications for the stakeholders there. In this section, we assess if there is a good match between the present destination area features, assets, potential and current markets. Moreover, we assess 'possible futures' for the destination area – the expected destination development over the next five years and possible future scenarios. Southbank (Chapter 2), Central Business District (Chapter 3) and Dockland (Chapter 4) in Melbourne are explored in terms of possible futures of tourism developments and their ramifications for stakeholders.

Part II: The Probable Future: Analyse and Evaluate: Match and Probable Future

Part II includes the probable future, what is most likely to happen at the destination and what will be the consequences for stakeholders? In this section, the emphasis is on the *probable future*, in which we identify certainties and uncertainties at the destination level, and identify the global drivers of change, changes in demographics, economic activity and consumer preferences and technology. The focus is on Penang, Malaysia, where we assess the match between the present destination area, features, assets, potentials and the current markets. Furthermore, we assess the 'probable future' (one future) of the destination area: what is most likely to happen (change analysis) over the next five years, and the consequences for the stakeholders. George Town (Chapter 5), Gurney Drive (Chapter 6) and Penang Hill (Chapter 7) are the locations that have been investigated.

Part III: The Preferable Future: Consult and Advice: Match and Preferable Future

Part III focuses on Chiang Mai, Thailand, where we discuss the numerous issues this destination faces and provide practical recommendations to the various players to assist them in attaining the destination's preferred future. In other words, we assess the current strategic position of the stakeholder groups based on the match between the present destination area, features, assets, potential and the current markets. Moreover, we define the most preferable development of the area over the next five years for the stakeholder group. Finally, we advise on effective implementable interventions required to achieve the destination area's preferable future in the three locations in Chiang Mai: Old Town (Chapter 8), Nimman area (Chapter 9) and (Chapter 10) Hmong Doi Pui.

Part IV: Conclusions by the Authors

The final chapter identifies common themes and issues from the preceding chapters and produces conclusions concerning the nature of, and implications to be drawn from, the adjustment of practices of future scenarios.

References

Amore, A. (forthcoming) (2024). Tourism, public policy and governance. In C. M. Hall (Ed.), *The Wiley Blackwell Companion to Tourism*. Wiley, Chichester, West Sussex, PO19 8SQ, UK.

Amore, A., & Adie, B. A. (2021). (Re) igniting tourism in cities after COVID: The same old risks of the "new normal". In V. Pecorelli (Ed.), *From Overtourism to Undertourism: Any Sustainable Scenarios in the Post Pandemic Time?* (pp. 13–31). Milan, Unicopli.
Amore, A., & Hall, C. M. (2017). National and urban public policy agenda in tourism. Towards the emergence of a hyperneoliberal script? *International Journal of Tourism Policy*, *7*(1), 4–22.
Bergman, A., Karlsson, J. Ch., & Axelsson, J. (2010). Truth claims and exploratory claims – An ontological typology of future studies. *Futures*, *42*(8), 857–865.
Brouder, P. (2020). Reset redux: Possible evolutionary pathways towards the transformation of tourism in a COVID-19 world. *Tourism Geographies*, *22*(3), 484–490.
Dodds, R., & Butler, R. W. (2008). Inaction more than action. In S. Gössling, C. M. Hall, & D. B. Weaver (Eds.), *Sustainable Tourism Futures: Perspectives on Systems, Restructuring and Innovations* (pp. 43–57). Routledge, New York.
Duffy, R. (2015). Nature-based tourism and neoliberalism: Concealing contradictions. *Tourism Geographies*, *17*(4), 529–543.
Dwyer, L. (2023). Resident well-being and sustainable tourism development: the 'capitals approach'. *Journal of Sustainable Tourism*, *31*(9), 2119–2135.
Dwyer, L. (2024). 'Measuring what matters': Resident well-being and the tourism policy cycle. *Sustainability*, *16*(22), 19.
Enger, A., Sandvik, K., & Iversen, E. (2015). Developing scenarios for the Norwegian travel industry 2025. *Journal of Tourism Futures*, *1*(1), 6–18.
Fontela, E., & Hingel, A. (1993). Scenarios on economic and social cohesion in Europe, *Futures*, *25*, 139–315.
Godet, J. (2000, September). GPS/GALILEO radio frequency compatibility analysis. In *Proceedings of the 13th International Technical Meeting of the Satellite Division of The Institute of Navigation (ION GPS 2000)* (pp. 1782–1790).
Gössling, S., Scott, D., & Hall, C. M. (2021). Pandemics, tourism and global change: A rapid assessment of COVID-19. *Journal of Sustainable Tourism*, *29*(1), 1–20.
Hall, C. M. (2008). *Tourism Planning: Policies, Processes and Relationships*. Pearson, London.
Hall, C. M., & Jenkins, J. M. (1995). *Tourism and Public Policy*. Routledge.
Hall, C. M., Scott, D., & Gössling, S. (2021). Pandemics, transformations and tourism. In C.M. Hall, D. Scott & S. Gössling. Be Careful what you wish for. In *Global Tourism and COVID-19* (pp. 123–144). Routledge.
Hall, C. M., & Seyfi, S. (2021). Covid-19 pandemic, tourism and degrowth. In C. M. Hall, L. Lundmark, & J. J. Zhang (Eds.), *Degrowth and Tourism: New Perspectives on Tourism Entrepreneurships, Destinations and Policy* (pp. 220–239). London, Routledge.
Hancock, T., & Bezold, C. (1994). Possible futures, preferable futures. *Health Care Forum*, *37*(2), 23–29.
Hartman, S. (2023). Destination governance in times of change: A complex adaptive systems perspectives to improve tourism destination development. *Journal of Tourism Futures*, *9*(2), 267–278.
Henchey, N. (1978). Making sense of future studies. *Alternatives*, *7*(2), 24–27.
Higgins-Desbiolles, F. (2020). Socialising tourism for social and ecological justice after COVID-19. *Tourism Geographies*, *22*(3), 610–623.
Higgins-Desbiolles, F. (2021). The "war over tourism": Challenges to sustainable tourism in the tourism academy after COVID-19. *Journal of Sustainable Tourism*, *29*(4), 551–569.
Hiltunen, E. (2009). Scenarios: Process and outcome. *Journal of Futures Studies*, *13*, 151–152.

Inayatullah, S. (2013). Futures studies: Theories and methods. *There's a Future: Visions for a Better World, 30(2024042020310148585438)*.
Joseph, C. F. (2000). Scenario planning. *Technological Forecasting and Social Change*, *65*, 115–123.
Kahn, H., & Wiener, A. J. (1967). *The Year 2000: A Framework for Speculation on the Next Thirty-Three Years*. New York: The Macmillan.
Kosow, H., & Gaßner, R. (2008). *Methods of Future and Scenario Analysis.* Wiesbaden, Germany, German Development Institute.
Lindgren, M., & Bandhold, H. (2003). *Scenario Planning*. London: Palgrave.
López-Rodríguez, M. D., Oteros-Rozas, E., Ruiz-Mallèn, I., March, H., Horeca-Milcu, A. I., Heras, M., Cebrián-Piqueras, M. A., Andrade, R., Lo, V. B., & Piñeiro, C. (2023). Visualizing stakeholder's willingness for collective action in participatory scenario planning. *Ecology and Society*, *28*(2), 5.
Major, J., & Clarke, D. (2022). Regenerative tourism in Aotearoa New Zealand – A new paradigm for the VUCA world. *Journal of Tourism Futures*, *8*(2), 194–199.
Malaska, P., Malmivirta, M., Meristö, T., & Hansen, S. O. (1984). Scenarios in Europe—who uses them and why?. *Long Range Planning*, *17*(5), 45–49.
Martino, J. P. (2003). A review of selected recent advances in technological forecasting. *Technological Forecasting and Social Change*, *70*, 719–773.
Morgan, N. (2012). Time for 'mindful' destination management and marketing. *Journal of Destination Marketing & Management*, *1*(1&2), 8–9.
Morrison, A. (2013). *Marketing and Managing Tourism Destinations*. London, Routledge.
Nyaupane, G., & Buzinde, C. (2017). Scenario planning: A planning tool for an uncertain future. Paper presented at the TTRA conference 2017, 20–22 June.
Oner, A. M. (2010). On theory building in foresight and future studies: A discussion note. *Futures*, *42*(October), 12. Available at www.sciencedirect.com.helicon.vuw.ac.nz/science/journal/00163287/42/9 (Accessed 10 December 2024).
Pechlaner, H., Zacher, D., Eckert, C., & Petersik, L. (2019). Joint responsibility and understanding of resilience from a DMO perspective – An analysis of different situations in Bavarian tourism destinations. *International Journal of Tourism Cities, 5*(2), 146–168.
Peeters, P., & Papp, B. (2023). *Envisioning tourism in 2023 and beyond. The changing shape of tourism in a decarbonising world.* Travel Foundations. Available at https://pure.buas.nl/ws/portalfiles/portal/27136592/Peeters_Papp_EnvisionTourism_report.pdf (Accessed 12 December 2024).
Poskitt, S., Waylen, K. A., & Ainslie, A. (2021) Applying pedagogical theories to understand learning in participatory scenario planning. *Futures*, *128*, 102710.
Postma, A., Hartman, S., & Yeoman, I. (2024). *Scenario Planning and Tourism Futures: Theory Building, Methodologies and Case Studies.* Brisntol: Channel View.
Rigby, D., & Bilodeau, B. (2007). Selecting management tools wisely. *Harvard Business Review*, *85*(12), 20–21.
Rubin, A., & Kaivo-Oja, J. (1999). Towards a futures-oriented sociology. *International Review of Sociology*, *9*(3), 349–371.
Schnaars, S. P. (1987). How to develop and use scenarios. *Long range planning*, *20*(1), 105–114.
Šegota, T., Mihalič, T., & Kuščer, K. (2017). The impact of residents' informedness and involvement on their perceptions of tourism impacts: The case of Bled. *Journal of Destination Marketing & Management*, *6*(3), 196–206.
Strand, S. (1999). Forecasting the future: Pitfalls in controlling for uncertainty. *Futures*, *31*, 333–350.

Torkington, K., Stanford, D., & Guiver, J. (2020). Discourse (s) of growth and sustainability in national tourism policy documents. *Journal of Sustainable Tourism*, *28*(7). 1041–1062.
Varum, C. A., & Melo, C. (2010). Directions in scenario planning literature – A review of the past decades. *Futures*, *42*(4), 355–369.
Yeoman, I., Davies, J., Wheatley, C., Mars, M., Schänzel, H., & Butteriel, S. (2012). *Tourism2050: Wellington, Scenarios for New Zealand.*

Part I

The Possible Futures

Collect and Assess: Match and Possible Futures

2 The Possible Futures of South Bank, Melbourne, Australia

Introduction

Southbank is an urban neighbourhood in Melbourne that fronts the Yarra River, which runs alongside 2 kilometres of the suburb. The front of Southbank has a wide pedestrian promenade that offers scenic views of the city skyline and the river. Southbank is approximately 1 kilometre away from Melbourne's Central Business District (CBD), and is bordered by Montague Street in the west and St. Kilda Road in the east. The Princess Bridge, Sandridge Bridge, Evan Walker Bridge and Queens Bridge are just some of the bridges that connect Southbank to Melbourne's CBD. They enable companies to settle on the Southern bank of the Yarra River and therefore expand the business environment into the Southbank area (Herald Sun, 2017). Southbank is one of Melbourne's most high-density residential suburbs, with 18,800 people (in 2016), of whom most call a multi-level apartment building home. The Eureka Tower, currently Melbourne's tallest building, is located in Southbank. This massive residential tower includes the Skydeck located on the 88th floor. This is Australia's highest viewing platform, providing spectacular views of the bay, city and surroundings during the day and night (TravelVictoria, 2020).

Southbank is also known for its cultural facilities which include the leading contemporary art gallery in Melbourne and The Australian Centre for Contemporary Art. Other cultural facilities include the Australian Ballet Centre, Victorian College of Arts, National Gallery of Victoria and Melbourne Recital Centre. These institutions are found in the Arts precinct of Southbank. Heading west is the Southgate shopping centre that offers cafés, restaurants and retail shops overlooking the river. Further west, the riverfront walkway is known as Yarra Promenade, which is home to The Crown Entertainment Complex that includes a five-star hotel, restaurants a cinema and the Crown Casino itself. The Melbourne Convention and Exhibition Centre is located at Spencer Street, where exhibitions and events take place. The riverfront extends west to the entrance into the South Wharf precinct.

The overall aim of this chapter is to assess the possible futures for Southbank in Melbourne, Australia, and the implications for the stakeholders involved. In this chapter, we assess if there is a good match between the present destination area features, assets, potential and the current markets. Moreover, we assess the 'possible futures' for the destination area – the expected destination development over

DOI: 10.4324/9781003570240-3

the next five years and possible future scenarios. The purpose is to provide insight into the micro- and small-size enterprises (MSEs) in the area of Southbank in terms of their current situation as well as future developments. (Enterprises are defined as legal entities that possess the right to be able to conduct business on their own. They are able to establish bank accounts for example, or own property, enter into contracts and/or incur liabilities.) These entities are subject to being either a non-profit organisation, a corporation, a quasi-corporation or an unincorporated enterprise (OECD, 2020). For the purpose of this research, we have defined MSEs as enterprises with a maximum of 50 employees on a contract.

Contextual Analysis

The Melburnian suburb of Southbank can be divided into several precincts. The suburb is located on the Southern side of the Yarra River, bordered by the river in the north and the highway M1 in the south (Stanton, 2018; TravelVictoria, 2020). South Wharf, the neighbouring casino area as well as the Southgate are precincts within this area which align on the waterfront, whereas the precincts of Boyd, Arts and Malthouse are situated behind the waterfront towards the town of St. Kilda.

For the research process, it was important to define the different precincts of the Southbank area in order to illustrate the differences and similarities between the areas. In Figure 2.1, the red line represents the boundaries identified for the study. While the blue line indicates the boundaries of Southbank by Google Maps. For the purpose of this study, one specific set of boundaries was utilised. Therefore, we merged the two for a complete representation of the area.

History of Southbank

Throughout its history, Southbank has experienced major development. Before the European settlement, the area consisted of low-lying swamps where a first nations tribe lived. By the middle of the 18th century, Southbank had become an industrial centre of Melbourne, being part of the city's first port with several factories, wharves and warehouses. In 1980, the Arts precinct also appeared on Southbank land. Facing the global recession in 1991/1992, urban renewal took place in the suburbs to stimulate economic growth, starting with the construction of the Southbank promenade as well as a pedestrian bridge. This development proved attractive to residents as well as tourists. This success also enabled the construction of the Esso headquarters, the Melbourne Convention and Exhibition Centre and the Casino as well as several residential buildings further west along the riverside. At this point, the Sandridge pedestrian bridge connects the northern and southern banks of the Yarra River, which led to many companies settling in Southbank because of the immediate connection to the CBD. Underneath the bridge, Ponyfish Island was established, which is a floating platform on the river (Southbank Residents Association, n.d.). At the beginning of 2008, the area of South Wharf

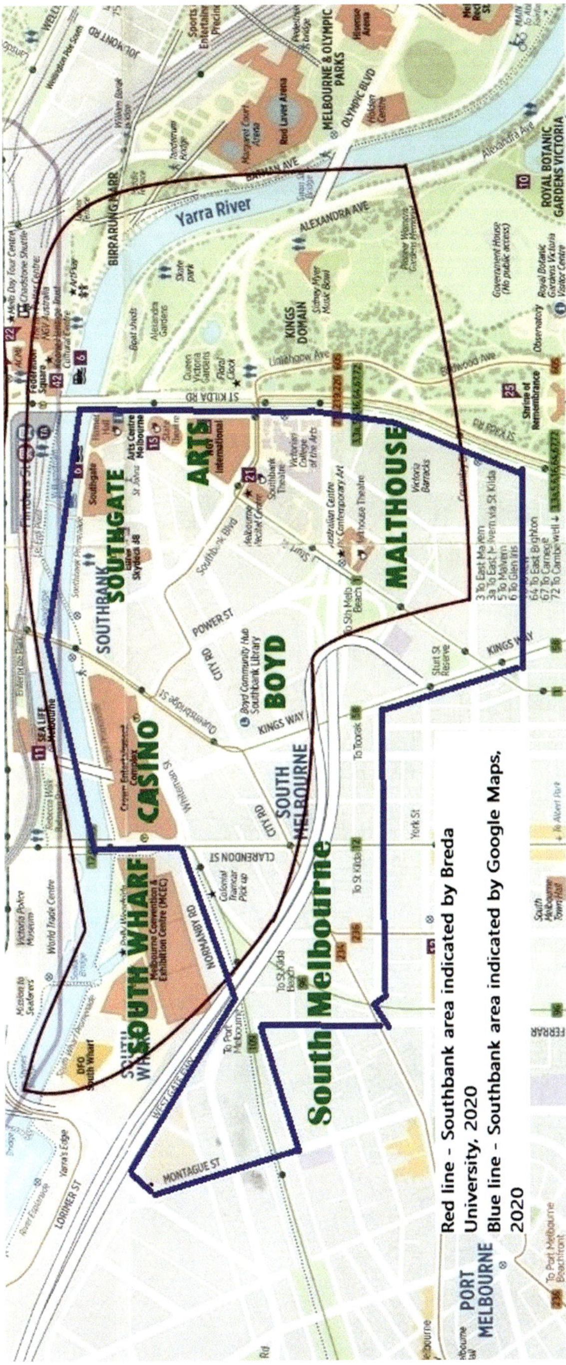

Figure 2.1 Combined map of Southbank. Map data ©2025 Google annotated by students on MA Tourism Destination Management, at Breda University of Applied Sciences, 2025, The Netherlands.

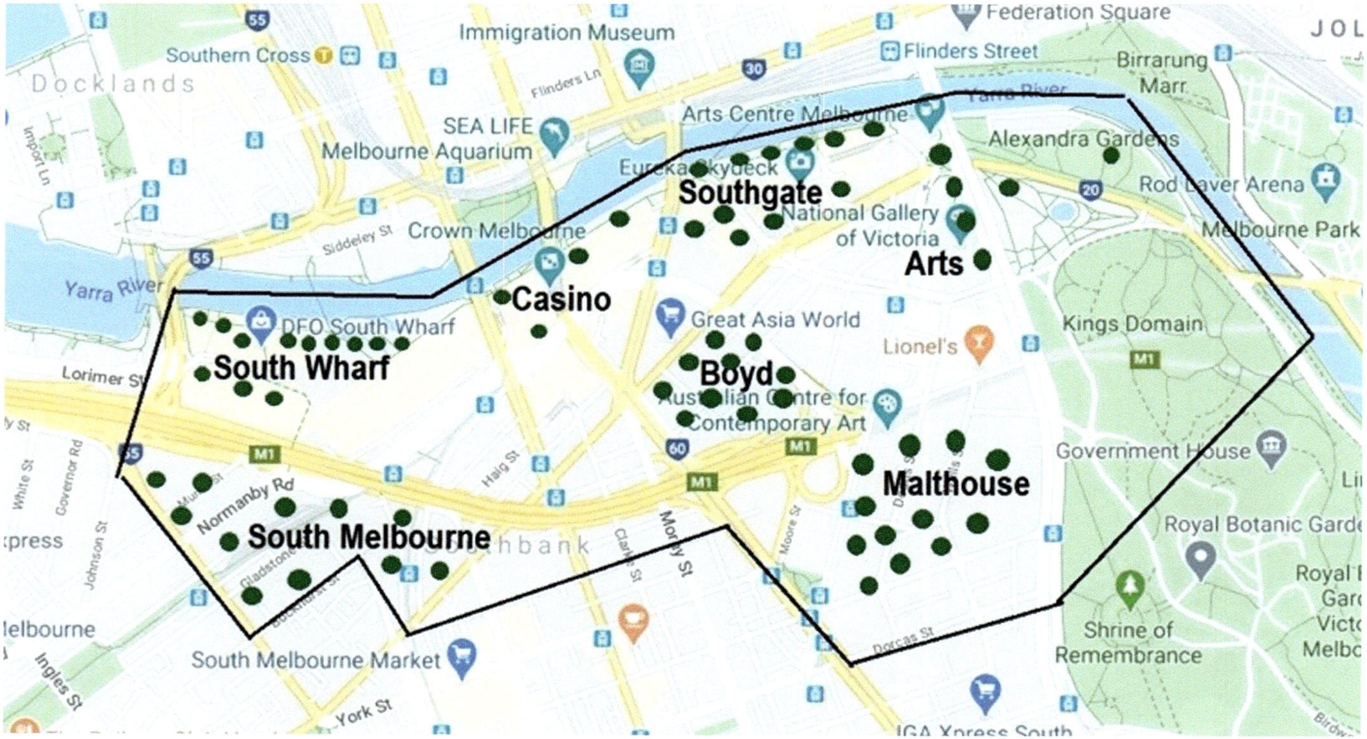

Figure 2.2 Representation of residents in Southbank. Map data ©2025 Google accessed via www.travelvictoria.com.au/southbank/maps/, annotated by students on MA Tourism Destination Management, at Breda University of Applied Sciences, 2025, The Netherlands.

was constructed as part of a 42 million AUD investment to add value to the area (City of Melbourne, 2020). See Figure 2.2 for the representation of residents in Southbank.

Current Developments in Southbank

Looking at the current situation, Southbank is known for its art scene and the high share of residential buildings in the area. For the future of the inner suburbs, the major goal is to transform towards a more green and creative environment (City of Melbourne, 2020). With 18,000 people living in the area, as well as 45,000 employees working there, this Melbournian suburb is the most densely populated one within the city. Additionally, there are 80,000 visitors coming to the neighbourhood every day (City of Melbourne, 2020).

Southbank functions as a major contributor to the economic success of the city, especially the CBD, and finds itself experiencing constant growth (Southbank News, 2020). This economic growth is leading to an extension of the CBD area across the Yarra River towards the south of Melbourne. However, the risk of creating so-called economic ghettos arises with the clustering of economic power only in a specific area of a city, leading to a power and economic imbalance (Herald Sun, 2017).

This creates the need to examine more closely the clustering of certain socio-economic profiles in the area of Southbank. The median age of the population ranges around 30 years, with most residents having received tertiary education. This leads to the assumption of a rather high-income structure. The average household in 2-bedroom apartments hosts 1.9 people, mostly couples with no children. In terms of heritage and nationality, there is a rather low percentage of Australian-born residents, 29 percent, in comparison with the 64 percent of Australians in the state of Victoria overall. In comparison, the proportion of Chinese people ranges around 13 percent in Southbank compared with 2.7 percent in Victoria. Of the working residents in Southbank, 20 percent each commute to work by car or tram (Qpzm, n.d.).

In terms of environmental issues, all of Melbourne and Southbank are battling flooding every year. The last major flood occurred in 2010. Every year, there arise smaller flooding issues in the area especially around Boyd along Kings Highway. This puts residents as well as businesses at risk regarding profit–loss and presents a definite inconvenience for the area. Southbank cannot be seen as a generic neighbourhood but rather as a diverse mixture of different areas, consisting of the Arts precinct, Boyd, Malthouse, South Wharf, the Casino area as well as Southgate.

While the Arts Centre characterises the variety of bars and restaurants in the area, various museums, as well as the meeting hall serve as a communal meeting point for residents to share their ideas and creations. Boyd represents itself as both residential and industrial (VisitVictoria, n.d.b), and with several high-rising buildings and closely located to the highway M1, it borders southern Melbourne. The Boyd community centre acts as a meeting point for the residents with an additional park functioning as a backyard for the area's residents (City of Melbourne, 2020a). In the back of the Southbank area is located the Malthouse precinct, honouring the brewery culture in that area. The Malthouse theatre can be seen as the central point in the precinct, illustrating innovation and revitalisation of old building structures as well as entertainment (VisitVictoria, n.d.-a).

Southgate precinct, being the commercial hub for shopping and dining experiences since 1992, is located at the waterfront. Its focus lies in the shopping centre of the same name, including a food court and various fashion shops (City of Melbourne, 2020b).

Moving further west on the waterfront, one comes across the Casino precinct, encompassing the Crown Entertainment Complex as one of the major attractions which can be seen as one of the nightlife centres in the city of Melbourne. Eight high-rising towers provide entertainment such as concerts, sports events as well as shopping, and the venue functions as a wedding venue. The complex aims to attract different customer segments by offering casual as well as fine dining. In terms of customers, a focus is on Asian customers (OnlyMelbourne, n.d.; Crown Melbourne, 2020). Finally, there is the South Wharf district, located in the most westerly part of Southbank with its promenade and cargo warehouse companies (City of Melbourne, 2019). Additionally, it is home to the DFO shopping centre offering shopping experiences (DFO South Wharf, n.d.).

Stakeholders in Southbank

The past, current and especially future developments in the Southbank area concern several stakeholder groups, within the MSEs as well as other outside interests. Tourists, residents and corporates are the customer groups for these specific businesses and are therefore considered important stakeholders. Furthermore, the City Council of Melbourne interacts both as a policymaker and a governing power with the businesses. Other businesses of a different size located in the same area also play a role.

Looking further into the stakeholder group of the MSEs and the different branches within, we distinguished the different businesses found in the Southbank area into the categories of tourism-related and non-tourism-related businesses. The visualisation of these different types of MSEs can be found in Figure 2.3.

Future Developments

There are certain developments and global drivers of change happening on a larger scale, which influence the business environment within Southbank. Examples include sustainable development and the concentration of economic power in certain areas of the world, as well as population growth and diversification which are issues affecting developments both in Melbourne and all of Australia (Australian Government, 2019; ACCA, n.d.). The fact that increasing numbers of people are moving to urban areas in Australia, accentuating economic imbalance within the country can also be seen as another reason for the establishments of these projects by the city council.

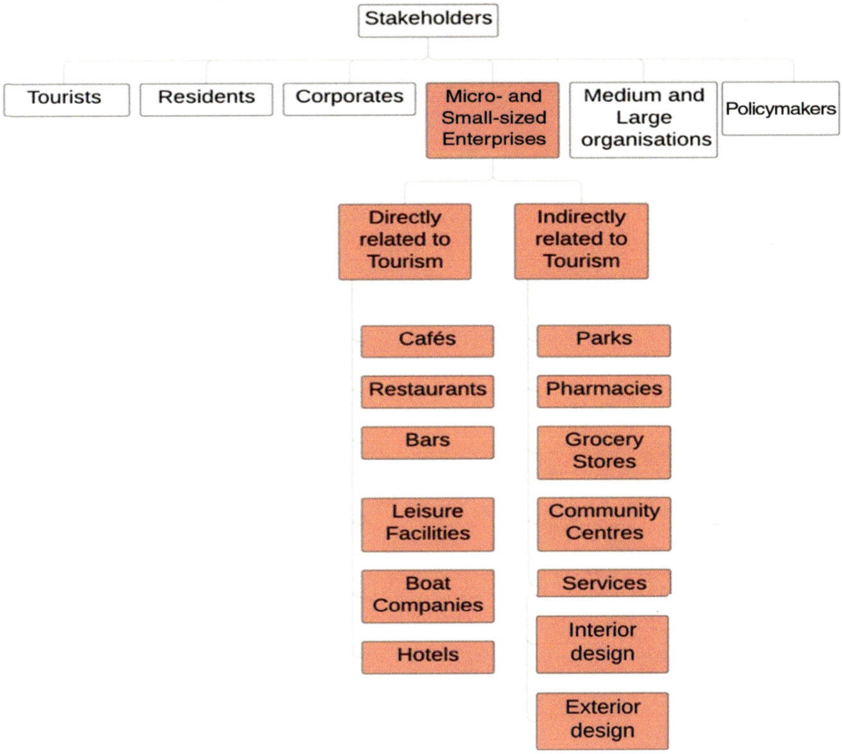

Figure 2.3 Stakeholders in Southbank.

Source: Created by the authors.

Additionally, there are local factors such as the strong growth in educational tourism in Australia, the changes wrought by the coronavirus and the increasing threat of bushfires which might affect MSEs within the Southbank area (Australian Government, 2019).

Looking into the future of this specific area of Melbourne, the goal is to 'transform Southbank from grey to green' (City of Melbourne, 2020, par. 3). In pursuit of this goal, there were several projects set in place, which were anticipated to have major influences on the development of the Southbank area. For example, the city council is planning an urban forest on City Road to help with issues such as flooding and to encourage the citizens to live a more active lifestyle by commuting to work by foot or bicycle, while at the same time cooling the local climate. There are plans in place for the further revitalisation of the Southbank along the Southgate promenade involving relaxation opportunities, additional lights and renewed stairs and ramps to make the environment more attractive to residents and visitors of the area, in conjunction with additional green spaces and parks throughout Southbank

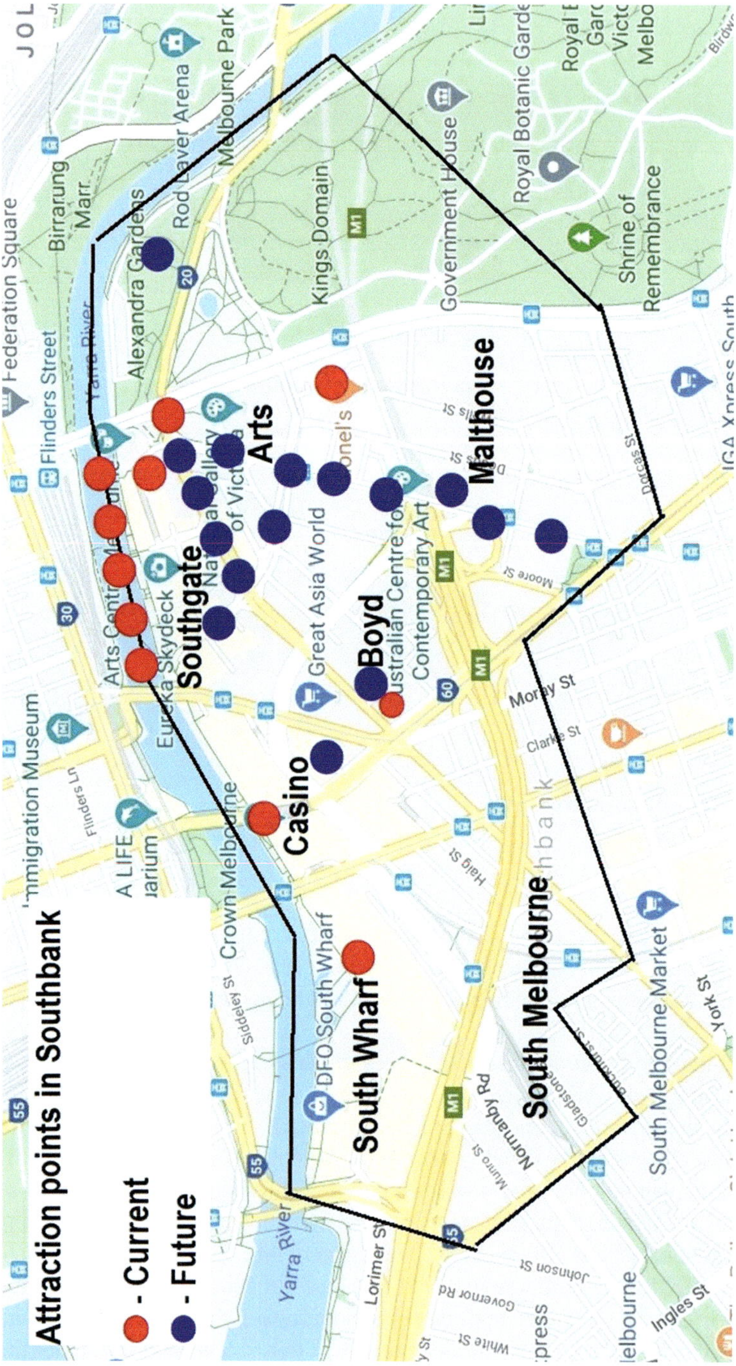

Figure 2.4 Current and future developments. Map data ©2025 Google accessed via www.travelvictoria.com.au/southbank/maps/, annotated by students on MA Tourism Destination Management, at Breda University of Applied Sciences, 2025, The Netherlands.

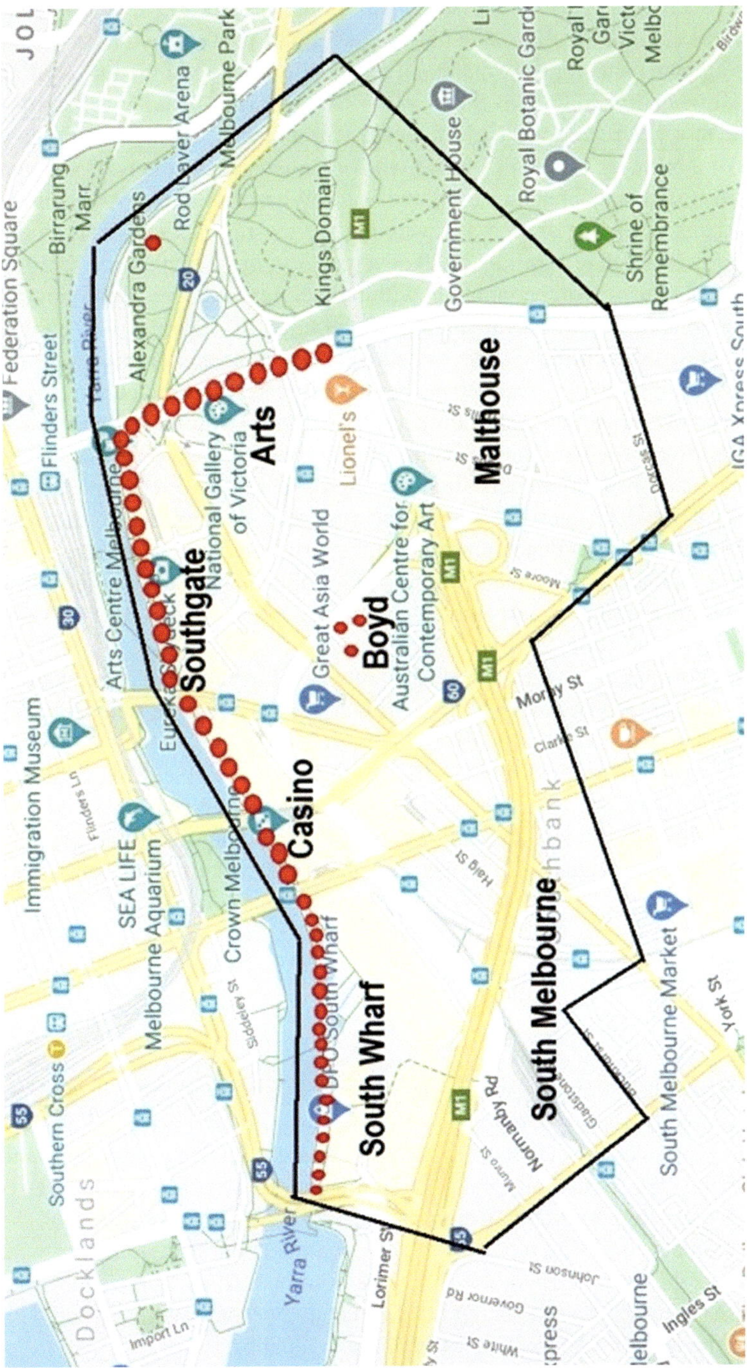

Figure 2.5 Representation of tourists in Southbank. Map data ©2025 Google accessed via www.travelvictoria.com.au/southbank/maps/, annotated by students on MA Tourism Destination Management, at Breda University of Applied Sciences, 2025, The Netherlands.

being a major focus of future city development. Besides the addition of various green spaces such as the parks along Southbank Boulevard, the city council is also investing in the creation of affordable housing, especially in the Boyd area.

A second major focus for the Southbank area is the creation of the image of an arts district, especially along Sturt Street in the Art precinct. The latter is recorded in Amendment C323 regarding the Melbourne Arts precinct, stating that the first four floors of a building can be decorated with so-called active street art. Another project for the Southbank area, including the South Wharf precinct, is Amendment C305 Southbank Heritage, aiming at reviewing the existing heritage sites in the area and giving heritage recognition to 20 new places (City of Melbourne, 2020b). To accommodate the growing population within the Southbank area, there are several constructions in the process including the building of Australia 108, aiming to be the highest tower in the Southern hemisphere that creates living space for residents (Multiplex, 2016). Another residential project is the Palladium Tower Apartment or the Normanby (Salvo Property, Normanby). Another project is the building of a tower near the Crown complex encompassing apartments as well as offices and a focus on entertainment similar to the Casino complex (Baume, 2019).

The Yarra River delineates one side of the area, and the Yarra River Business Association coordinates its development with the CBD area on the northern riverside and the Southbank area on the lower riverside. Other ideas are the usage of the river itself as a means of transport and the greening of the Sandridge Bridge (Yarra River Business Association, 2019).

Figure 2.4 portrays the current as well as future developments and constructions in the Southbank area more precisely. This provides the basis for the analysis of the possible future of MSEs in the suburb.

These changes in the environment of the businesses will clearly influence the MSEs in the area and thus these topics were engaged with in the interviews with the respective businesses to explore their perceived impacts on current and future developments. Figure 2.5 shows also the representations of tourists in Southbank area.

Current Situation and Possible Futures

Arts Precinct

The Arts precinct has been situated in the Southbank area since 1980, behind the Southgate shopping centre, bordering St. Kilda Road and the Botanical Garden (Southbank Residents Association, n.d.). With a large number of theatres and museums such as the Australian Centre for Contemporary Art in the area, the Arts precinct represents an important element of the Southbank area. Additionally, the precinct represents a meeting point for the creative minds of the community with venues such as the testing grounds (City of Melbourne, 2020a). In the future, the city council is planning on renovating Sturt Street and decorating the facade with street art on the first four floors (City of Melbourne, 2020b).

To determine the strength of the current market as well as the potential future development of the MSEs specifically within the Arts precinct, the research team examined businesses within the food and beverage branch, a community-based pavilion and a skate park within the Botanical Garden, as well as an organic grocery store located in the more residential area behind the Arts Centre. These MSEs represented branches of the major business types in the area enabling a diverse impression.

Looking at the current customer segments, it is clear that the precinct is an area frequently visited by Melburnians. People working at the Arts Centre, corporates who are passing the precinct on the commute to work, residents living in the area as well as university students are present. There are also tourists coming to the area, mostly because of interest in the multiple art performances. To cover the interests and needs of these different customer groups, MSE in the Arts precinct are found primarily close to St. Kilda Road and the Royal Botanic Gardens as well as behind the art venues, with three bars at the Arts Centre closely connected to the venue itself, having a competitive advantage over similar businesses. With the current performance at the Arts Centre Melbourne being the musical *Chicago*, the Arts Centre has constructed two temporary bars to portray the background and time in which the storyline of the musical takes place. These bars mostly cater to visitors to the Arts Centre, both residents and tourists. However, the businesses connected to the Arts Centre have slightly higher prices compared with similar businesses in the rest of the Arts precinct. The businesses located within the immediate proximity such as the café in the university building and the mobile businesses in the container cater mostly to university students, and also employees from the surrounding institutions and art-interested tourists and residents passing by. They pose a cheaper alternative to the bars and cafés within the venue.

Additionally, there was also a skating park located within the Botanical Garden, encompassing a skating school as well as a small coffee seller. Interviews showed that this place is another tourist attraction in the area, hosting a variety of tourists from China, the United Kingdom and Italy. Many repeat visitors were found to be tourists, revisiting the park regularly, along with several residents.

The MPavillion across the road from the Arts Centre functions mainly as a meeting point for locals to communicate and share creative ideas. The not-for-profit policy of this specific business makes it distinct from other businesses. Interviews showed that the project belongs to the Naomi Milgrom Foundation, celebrating different architects each year and focusing on reinvestment in the community. It is frequented by all kinds of age groups within the local community, again depending on the content of the hosted events. Workers from the Arts Centre as well as university students in the art field were specifically mentioned to be the customers.

Looking at the general allocation of customer groups with a special focus on tourists, the further away from St. Kilda Road and the Botanical Garden, the fewer tourists seem to frequent the area. The back area towards the Boyd precinct presents itself as a predominantly residential area with fewer businesses located in the area. However, an organic grocery store located in the area behind the Arts Centre offers products at a higher price while being one of the few grocery stores in the area. The majority of MSEs located around the museums, theatres and arts venues along St. Kilda Road and the Botanical Garden profit from the touristic attractions along the road and the employees and tourists coming there, as well as the daily commute from the suburbs into the CBD area. Aspects such as community support and art are clearly emphasised in this part of the street. However, the rear of the Arts precinct seems to be a less attractive area for the businesses, since it consists mostly of stage doors and backs of the venues at St. Kilda Road with fewer businesses located in this area.

Following the past motto 'If you build it, they will come,' the city of Melbourne aims to redesign Sturt Street in the future (City of Melbourne, 2016). Being behind St Kilda Road, this area is expected to be more decorated with art and to become more welcoming to pedestrians. Another goal is to reuse older buildings and land by locating other cultural institutions in these spaces. This development is expected to widen the attractive areas of the Arts precinct further to the area behind St. Kilda Road. Consequently, the MSEs in the food and beverage branch along St. Kilda Road could expect an increase in competition with similar businesses allocating around Sturt Street as well.

Secondary research has shown that educational tourism in Australia is growing strongly, and enrolment is increasing in universities and educational facilities along St. Kilda Road (Australian Government, 2019). Consequently, this could result in an increase in profit for the MSEs catering especially students in this area.

Summing up, the Arts precinct is expected to extend further back behind St. Kilda Road and to have more businesses there which cater to art-interested people, both residents and tourists, with the larger hare being university students in the creative field and those working for the Arts Centre. As a result, the current businesses, especially the higher priced ones directly connected to the Arts Centre, might need to emphasise their unique selling point such as the connection to the background of the performances to withstand this increased competition. Businesses that are more community based with a not-for-profit approach such as the skate park or the MPavillion are not expected to be highly impacted by the developments of Sturt Street since they stand out through their outdoor activities and larger green spaces. However, global developments such as climate change and resulting bushfires have an impact on these outdoor facilities, leading to a decrease in visitors and consequently loss in profit. However, research and interviews showed that these events surprisingly came rather unexpected to the business-owners and were unprepared for.

Boyd Precinct

The precinct of Boyd is characterised by its residential atmosphere and the increased construction of newer further residential high-rising towers as well as the proximity to the highway M1 (VisitVictoria, n.d.b). Reflecting on this environment, the Boyd Community Hub was established, functioning as a meeting point for the locals (City of Melbourne, 2020a). There is a plan to increase green spaces as well as an urban forest to work against flooding issues in the future. Additionally, the construction of housing for residents on lower incomes aims to attract different kinds of residents to the area (City of Melbourne, 2020b).

Looking at the MSEs in the area, our research showed the presence of some cafés and restaurants, the mentioned community hub, as well as an Asian supermarket. The cafés and restaurants in the area cover continental and Asian cuisine. The surroundings include many residential towers with property offices or mostly empty office spaces on the lower floors. Paired with the proximity to the highway, the business environment seems to be less welcoming to tourists or residents at this point with a lot of noise and many disruptions in the area. This matches statements by the Yarra River Association that there were no residential amenities in the Boyd area in the past, these are only now starting to be built. The few older buildings in the area stand out by virtue of their appearance (often poorly maintained and under repair or used as advertisements) in contrast with newly built properties. However, these older buildings are mostly not well-maintained and are mostly covered in advertisements or construction scaffolds.

Most people simply pass by the area and its businesses, mainly commuting to or from work. This confirms the information received during interviews that the majority of (repeat) customers are indeed residents and the main source of income for the businesses, all well-established in the area, having been there for several years. An interesting development was discovered specifically regarding the cafés and restaurants in the lower-price segment in Boyd. There was an increase in frequentation by construction workers. These add another customer segment for the businesses, in contrast with the Asian grocery, which is mainly frequented by locals, among them many Asian students, especially in the evening hours, the latter making up the majority of repeat customers. This small enterprise covers the higher price segment due to the need to import many products and targets a more specialised market of Asian products.

Lastly, the community centre of Boyd encompasses a café as well as sports facilities and a green area with a playground around the building, a library and an assembly hall. This not-for-profit business model can be seen as the future vision for the area of Boyd according to policymakers. Observations showed that this facility is visited by people of all age groups, especially families with children, who make use of the community park, one of the very few in the area. The café within the community centre offers products at a lower price and therefore makes its services accessible for most customers, fitting the philosophy of an inclusive community. All the questioned businesses claimed to do well in terms of revenue and that the construction does not impact their business in any way. On the contrary, some of the businesses were even in the process of hiring new staff due to an increased frequentation of their business. The businesses perceive the new developments and constructions around them as an advantage since this means more residents coming in the future. However, our analysis gave us a different impression.

While the businesses in the area state that the construction workers temporarily make up a large share of their customer group, they do not seem to be preparing for the fact that once the developments are finished, this share will disappear. This will leave them with a loss in profit at least in the short-term that might not be immediately replaced by new residents moving to the area. A strong marketing approach will, therefore, be needed to advertise the businesses to the residents. Developers assume new residents to be moving to the area once the construction is done, however, it will take some time for word of mouth to spread the upgrades in the Boyd precinct. Consequently, this leaves the businesses with a certain downturn in terms of profit in the short term. However, in the long run, it can be assumed that the attractiveness of the area will be significantly increased, leading to more residents in the area and a more flourishing business environment clustering around the revitalised housing area. The share of corporates as a customer group is expected to increase due to the number of office spaces available in the area. Community-based models such as the community hub will also experience an increase in customers once the area experiences an influx of residents. Overall, the area of Boyd and its connected business environment will flourish in the long term. However, a slight decrease in profit is expected and some difficulties regarding the marketing of the developments and the new image of the area to the newer residents, who will be a key customer group for the success of the businesses.

Malthouse Precinct

The Malthouse emphasises the revitalisation of older buildings in the area and the honouring of the brewery culture with the Malthouse Theatre as a central attraction point (VisitVictoria, n.d.a). In terms of future plans, the revitalisation of Sturt Street also plays a role in this context. Since the street stretches to Malthouse, its art decoration is expected to also influence the Malthouse precinct (City of Melbourne, 2020b). Our research showed that Malthouse is both a highly residential and industrial area. However, there were very few grocery stores, restaurants, cafés and many constructions underway. The few businesses in the food and beverage branch in the area are MSEs offering products for relatively low prices with vegetarian and gluten-free options. Because of the construction in the area, the construction workers represent an additional customer segment for the business and bring in revenue. As a negative consequence, the construction sites are disrupting businesses that are nearby and might make it less attractive to other customers.

After researching the precinct of Malthouse we concluded that the environment is well-balanced, quiet and surrounded by office buildings, albeit with a large group of construction workers present compared with other precincts, as well as corporates and residents. Policymakers would like to extend the tourism area existing along the waterfront and the St. Kilda Road to the suburbs by redeveloping Sturt Street. This project will be to pedestrianize Sturt Street, which would allow the existing art venues to increase their outdoor spaces and green areas which in turn will add attractiveness to the venue. This will cluster all art-related businesses in one specific street and possibly extend the characteristics of the Arts precinct to Malthouse. In the future, once construction is finalised businesses may suffer because the construction workers will leave. After Sturt Street is restored and tourism is 'extended into the suburbs' however, Malthouse theatre will be one of the attraction points. This area will become a more mixed area like other precincts in Southbank with corporates, and residents will become more tourist-related. This could lead to new businesses relocating to the area, but we believe that these new businesses will not take away customers from existing businesses in Malthouse because the tourists will represent an additional customer segment to the area.

Southgate Precinct

Southgate Melbourne is a distinctive recreational area located on the Yarra River and was opened in 1992. It is a lively area with many restaurants and shops and it has always been a meeting place. As a result, Southgate's urban planning is intended to promote exploration and discovery of the area, and several artists show and sell their work along the promenade because of the number of tourists walking along this specific street. The heart of Southgate is directly opposite Flinders Street Station, where the Evan Walker Bridge connects the CBD with Southbank. In the Southgate area, there is a varied composition of restaurants, cafés and bars, including a food court and various retailers ranging from fashion, accessories, household items to daily service needs. The area is equipped with secure covered parking garages via City Road and Southgate Avenue, and Southgate is also accessible via Flinders station to alternate types of transportation (Southgate, n.d.a, n.d.b).

The research in Southgate revealed businesses in the food and beverage branch, furniture stores, a pharmacy, an Asian grocery store and a boat company. The MSEs in the area are a mix of companies dependent on residents, corporates or tourists. The information obtained from the interviews with businesses on the river indicated that the area around the river is the most touristic part of Southgate. Over the years, it has become increasingly busy there and pop-up restaurants are appearing. The interviewees mentioned that the summer is the peak season, which has a major impact on the businesses. The area adjoining the river attracts tourists, leading to a clustering of businesses catering to this customer group close to the Yarra River. This is also indicated by the street artists who can be found there who make and sell their art on the promenade. The restaurants, bars and cafés located in this area are focused on dining and offer no delivery service or take-away options. Given that food delivery is an important trend in Melbourne and has a 35 percent share of the food market, local businesses should consider offering this service to fit better into current trends (Maus, 2020).

Companies located outside the area of the Yarra River are heavily frequented by corporates, construction workers and residents, with an enormous walk-through of people, especially during office hours. These food and beverage businesses are very busy with mainly take-away beverages as well as breakfast and lunch. Remarkably, many companies open very early at 6 am, while closing around 3 pm. They are also closed on weekends, because of their focus on corporations. There are two Asian supermarkets, one located in Boyd and one in Southgate, outside the Yarra River area.

The food and beverage businesses outside the river area indicated that their businesses mostly rely on repeat customers, and corporates are by far the largest customer group followed by residents. An important development that has changed the business activity in Southgate has been the availability of the internet. An independent entrepreneur, who has had his business in the Southgate shopping mall for ten years, mentioned that the variety of stores has changed enormously during that period. Where previously there were clothing stores, bookstores, CD stores and flower shops, these have now disappeared or changed into food and beverage businesses. This development is because many people nowadays make purchases online, making it more difficult for non-food businesses to survive. (The same development is also taking place in other parts of Melbourne such as Chapel Street and Richmond; Maus, 2020). As a result, there are too many food and beverage businesses concentrated in this area of Southgate, which leads to severe competition and saturation of the market.

Overall Southgate is becoming increasingly tourist focused and busier, which has led to a rise in rents in the area. The government has tried to compensate for this with subsidies but this has not been sufficient according to the interviewees. The businesses, therefore, have compensated for the higher rent with downsizing in terms of staffing by relying on fewer or part-time staff. In general, for all businesses, the rent is rising. A risk for the MSEs is that their property might be taken over by larger businesses if they do not manage to cover their costs. This matches the view of the Yarra River Association, who stated that, for example, the small boating companies will eventually be taken over by larger organisations. Another development increasing this risk for the MSEs is major events within Melbourne such as the Australian Open at the Rod Laver Stadium, which attract tourists to other areas of the city. Tourists go directly to the arena and purchase goods and services there instead of utilising the local tourist hubs. Nonetheless, the future developments for the promenade including upgrades on outdoor areas with seating options and more lights will make it livelier.

Construction workers are a very important customer group in the area behind the waterfront but the income from them is only temporary. As soon as construction takes place in the neighbourhood, there is an increase in the number of construction workers as customers, but when construction is completed it has a major negative economic impact on the food and beverage sector for two reasons. First, the large group of customers from the construction sector will disappear. Second, when the construction is completed, more businesses will enter the food and beverage industry, leading to increased competition and oversaturation of the market. Given the limited diversity, these businesses must distinguish themselves clearly from others in terms of their business model and their product portfolio. One possibility for these MSEs to improve their market share is to adapt their trading hours to also include tourists and residents during the weekend as they are now closed on weekends.

Casino Precinct

The precinct encompassing the Crown Casino complex is characterised by a variety of entertainment options as well as higher-priced dining with a large spectrum of customer groups. With 22 million visitors in the year 2019, the Crown can be seen as a major attraction point in the Southbank area (Crown Melbourne, 2020). There was not a lot of information available regarding the future plans of this precinct due to it being a private organisation. However, it is noticeable that the competition will increase within the next few years because of the investments of Beulah International Ltd. in the 'Twisty Towers,' located close to the casino (Baume, 2019). Here again, the MSEs are clustered around the major attraction point of the Crown Casino, which attracts many Asian customers. The Crown states that Asian customers demand upscale venues, which makes the Asian market one of the most important market segments for the Crown (Crown Melbourne, 2020). The Crown Complex currently offers different price segments to a variety of customer groups, but with a focus on the higher-price Asian market segment becoming reliant on this group. Primary research in cooperation with another research group covering the medium and large enterprises in the Southbank area confirmed that the businesses located within the Crown complex have taken advantage of the casino being a major attraction point by clustering in its immediate surroundings. In the possible future of this precinct, assuming that the businesses there do not change their focus to other markets, the businesses within the Crown could be impacted by the key driver of change such as a decline in the Chinese market (Australian Government, 2019). This would mean a significant loss of a large share of their customer group not only to the casino but also to the surrounding businesses. Adding to the competition is the future project of the Twisty Towers across the street, which will make the market highly competitive in terms of accommodations and event venues. It will, therefore, be necessary for the Crown Complex to come up with a Unique Selling Point enabling the casino activity to stand out from similar venues and to secure not only its profit and that of the smaller businesses around it, otherwise, the latter might relocate closer to the new competition.

South Melbourne Precinct

The area located south of the M1 motorway is also considered a part of Southbank. This precinct was found to have both residential and industrial characteristics, as shown by the high number of car repair garages in one area and the community facilities such as a kindergarten or a primary school in another. The further from South Wharf in this precinct, the more two-level buildings are encountered with small businesses, both as part of the residential house and standing separately.

The places observed and in which interviews were conducted included the food and beverage and accommodation and service-related sectors as well as interior and exterior design. Most of the businesses are family-owned and came into existence around 10 years ago, meaning that they are well-established in the area. They offer products within the average price range. It represents a relatively quiet area with the car garages mostly frequented by residents.

The common theme was that the interviewees recognised Southbank as a developing and expanding area. There was a remark that these constructions and developments of new buildings alter the character of the whole area.

These developments also have an impact on the customer segments (residents and corporates) of the MSEs. The companies face problems in terms of accessibility as well as noise and a lack of parking spaces for their customers, which have negative consequences on the customer's ability to reach the business. The businesses in the branch of interior and exterior design in particular would benefit from the construction of new residential buildings. However, the MSEs in this area gave the impression that the developers care only about their economic benefit and not about social and community aspects.

Overall, the South Melbourne area is currently mostly an industrial area gradually being changed to mostly residential. The current customer segment for this area is mostly residents, especially for the car repair stations. There are two small hotels in the area which do not look busy suggesting a low number of tourists in this area.

The rather low prices in the area as well as the limited opening hours focused on the daytime market support the statement that this area is primarily industrial rather than a residential one. The future of this precinct can be predicted based on the happenings in the whole Southbank area, and it will get more residential and expand beyond the Motorway. In the future, the area will become more welcoming in terms of its business environment for the cafés and restaurants. Small businesses in this area, therefore, have the potential to grow due to an increase in residents moving in. Considering the current developments of digitalisation, it would be advisable for the MSEs to focus on online marketing strategies. A strong marketing approach will, therefore, be needed to advertise the businesses to the residents.

South Wharf Precinct

The South Wharf precinct was constructed at the beginning of 2008 as a part of a 42 million AUD investment to add value to the area located around the Yarra River. Currently, one can observe an entire promenade full of dining venues, cafés and bars – most of which are located in the restored 19th-century cargo sheds. A prominent attraction point in South Wharf is the Melbourne Convention and Exhibition Centre, which organises over 700 conferences and events per year. Another well-known site is the South Wharf DFO, a popular shopping centre. In terms of future developments in South Wharf (Amendment C305), Southbank Heritage aims at protecting the buildings and heritage located at the Western waterfront.

The research revealed mainly cafés, bars and restaurants as well as a massage salon established in the area around the Exhibition and Convention Centre. The MSEs indicated corporates as a major customer group during the week. The offering of a happy hour, as well as a workers' lunch in the food and beverage operations along the waterfront, supported this statement. Besides this type of customer, businesses also noted residents as repetitive customers and tourists as mostly one-time visitors adding a smaller share of the profit. These customer groups originate from the surrounding shopping centre and Marvel stadium, which hosts sports events. A statement by the Yarra River Association, supported the impression that the DFO shopping centre was initially not meant to be a tourist attraction, although it developed into a tourist magnet, especially for Asian customers over time.

All businesses recognised Southbank as being a developing area and specifically pointed towards the growing popularity of the Convention Centre and consequently the growing demand from corporates towards their businesses. However, the businesses are subject to seasonality with a clear peak in the summer and downtime in the winter. This, and the high dependency on the Convention Centre increases the risk to businesses in terms of potential profit loss if they focus heavily on just one customer group. Looking at the possible future of the South Wharf precinct, a decrease in the success of the MSEs located near the Convention Centre is not expected since the latter is regularly booked. On the contrary, the businesses themselves have experienced an up rise in customers over the last few years, which is expected to continue in the future.

Maps of the Current Situation

The distribution of the different customer segments across the respective precincts in Southbank as explained before is portrayed in the two maps (Figures 2.6 and 2.7). The dots indicate the frequentation corporations and construction workers in the specific areas. The last map (Figure 2.8) sums up the frequentation of the different customer groups in shares for each precinct to portray them in comparison with each other.

Overall Possible Future

The research has shown the general tendency of MSEs to cluster around major attraction points. These encompass the characteristics of green places and art, which are important aspects that policymakers wish to see represented in Southbank. However, currently, Southbank lacks these characteristics in various areas and precincts. Most of the characteristics are represented only at the riverfront, which consequently represents the tourist centre of the suburb and attracts tourism-related businesses in particular.

Away from the tourist hotspots, there is a domination of newly constructed buildings, which seem to lack character in terms of architectural detail and art components as well as missing greenery on the facades and recreational spaces. This currently makes the surroundings distant from touristic attraction points rather unwelcoming to tourists and does not match the intended image of the Southbank area. The future developments in Southbank point towards the expansion of these characteristics with new attraction points towards the back of the suburb of Southgate as shown in Figure 2.9. An increase of art and greenery on the grounds as well as on the facades is expected to add character to the surrounding precincts.

The mutual dependency of MSEs on a customer group is expected to highly influence the future development of the Southbank area. In the new tourist areas, there is a gradual new development of attraction points because of the construction and developments. Consequently, the allocation of new businesses and the

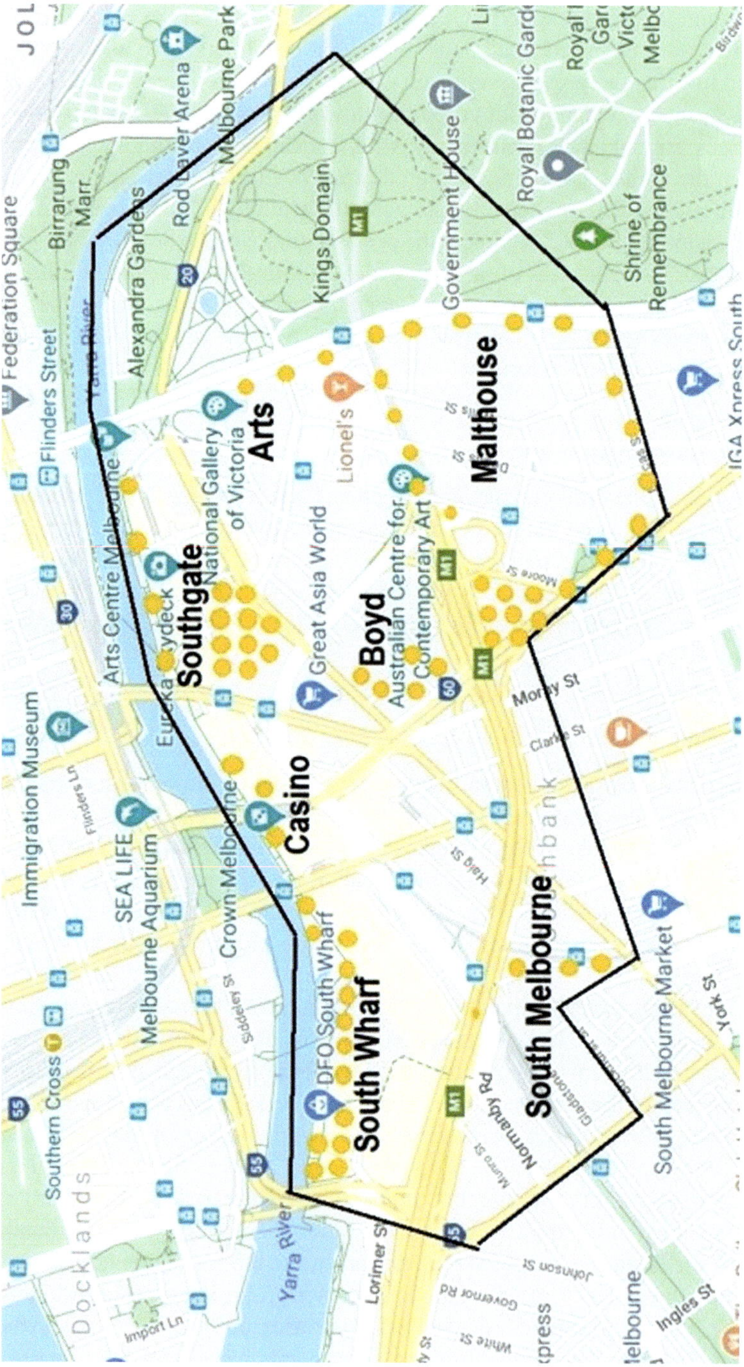

Figure 2.6 Representation of corporates in Southbank. Map data ©2025 Google accessed via www.travelvictoria.com.au/southbank/maps/, annotated by students on MA Tourism Destination Management, at Breda University of Applied Sciences, 2025, The Netherlands.

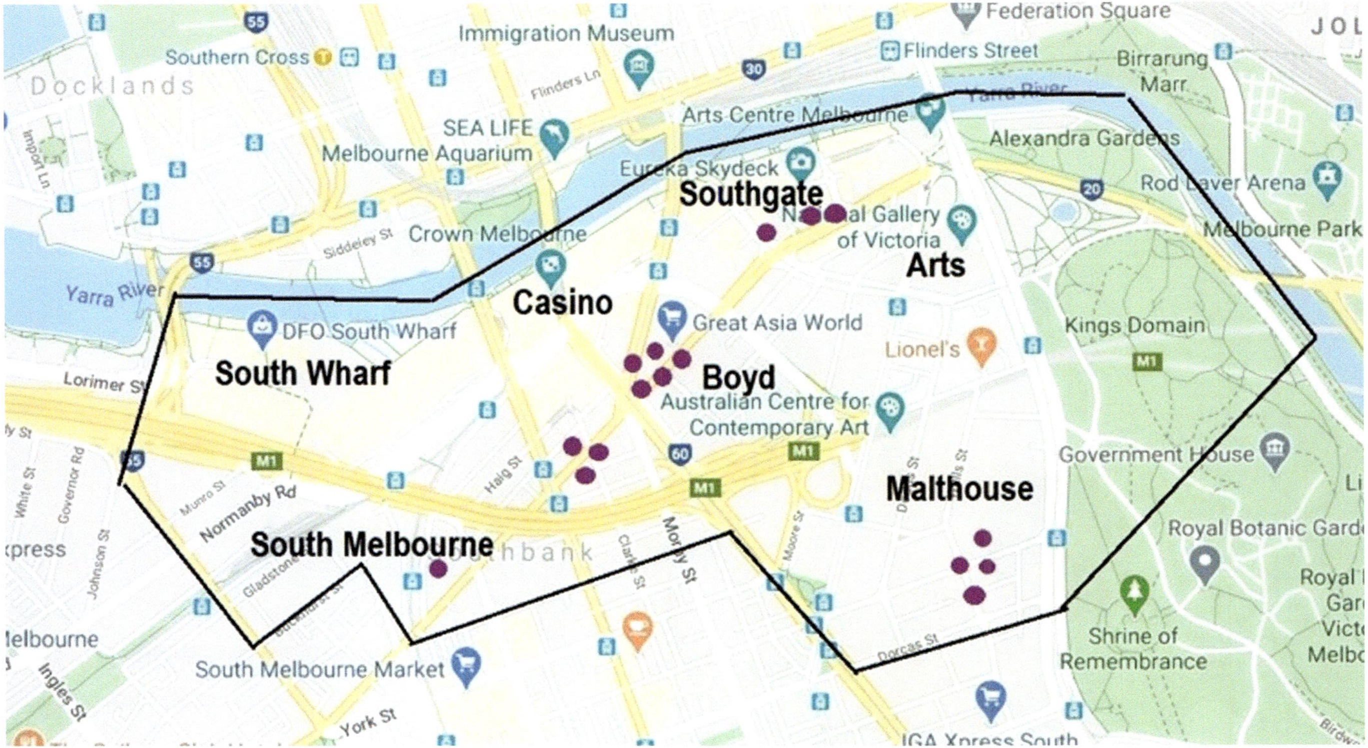

Figure 2.7 Representation of construction workers in Southbank. Map data ©2025 Google accessed via www.travelvictoria.com.au/southbank/maps/, annotated by students on MA Tourism Destination Management, at Breda University of Applied Sciences, 2025, The Netherlands.

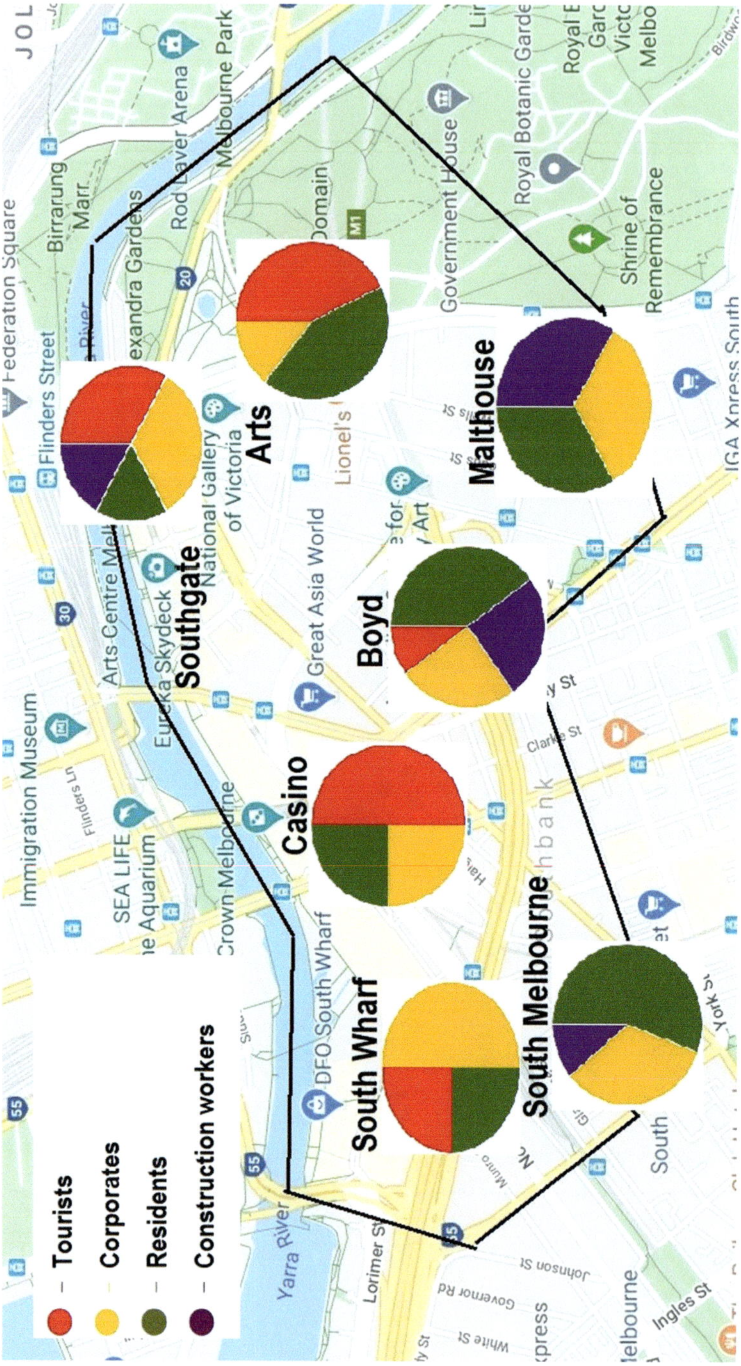

Figure 2.8 Diagrams displaying the shares of each customer segment in each precinct. Map data ©2025 Google accessed via www.travelvictoria.com.au/southbank/maps/, annotated by students on MA Tourism Destination Management, at Breda University of Applied Sciences, 2025, The Netherlands.

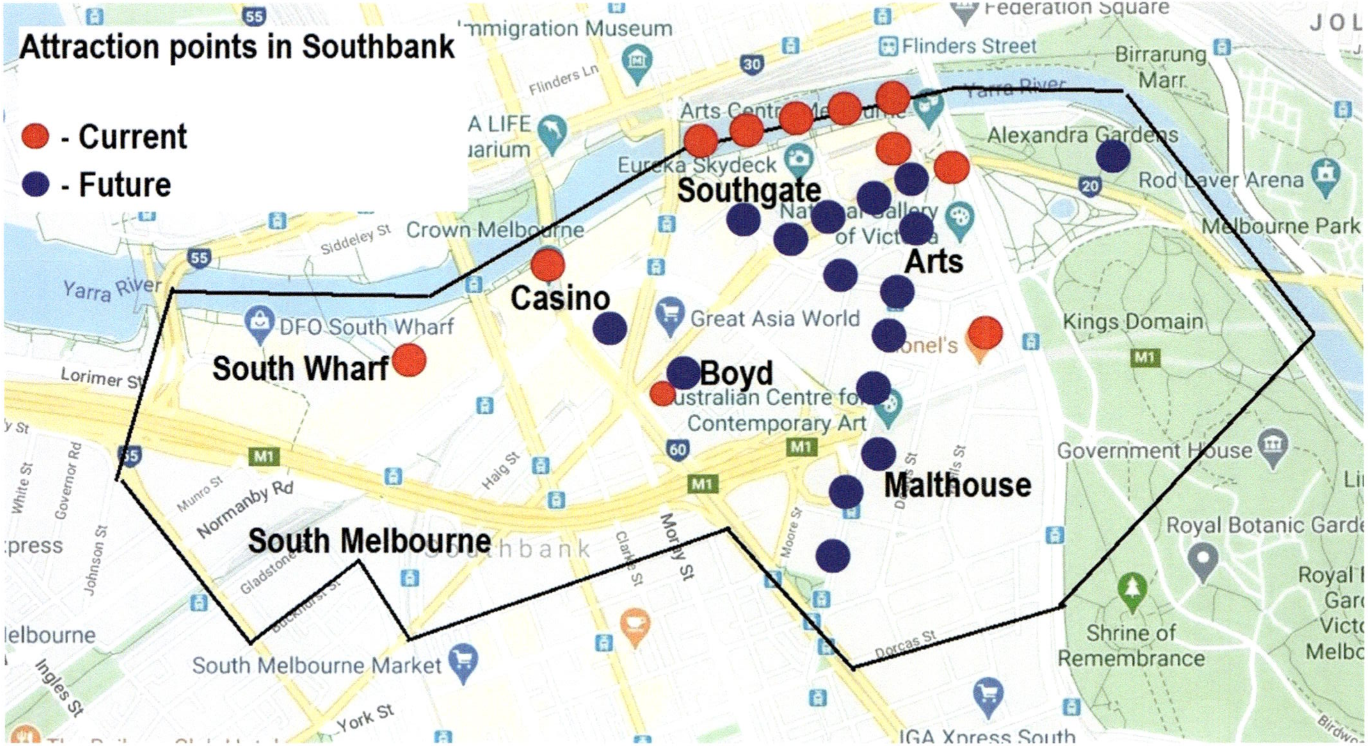

Figure 2.9 Representation of the current and future attraction points. Map data ©2025 Google accessed via www.travelvictoria.com.au/southbank/maps/, annotated by students on MA Tourism Destination Management, at Breda University of Applied Sciences, 2025, The Netherlands.

relocation of businesses in other locations to these areas is expected. Tourism-related businesses have been found to cater to and depend on travellers coming into the area. Therefore, tourism-related businesses, such as the food and beverage branch, are expected to settle in areas with an increasing tourism potential, such as that related to art and green spaces, mainly Boyd and Malthouse. Non-tourism-related businesses are consequently expected to settle in residential areas, being more frequented by corporates and residents, particularly South Melbourne and some parts of Boyd.

This relocation of certain businesses might solve some problems in the currently highly frequented tourist areas by diversifying and dispersing the mix of businesses, for example on the waterfront. The saturated market in the food and beverage branch could be spread out towards the other areas. Dispersal of the tourist masses is expected to relieve problems such as overcrowding of the Southbank promenade and the connected cyclist paths. Nevertheless, the spreading of the tourists will have significant consequences for those businesses dependent on the tourist customer group as well as to their cluster or attraction point. Businesses such as the boating companies rely on the river as the tourist attraction. If this attraction diminishes, these MSEs could be expected to disappear from locations such as the waterfront and to be replaced by larger companies. Consequently, the business landscape with respect to the MSEs will shift towards the back area of Southbank, with a slight division depending on a tourism or resident-related focus of the individual business.

References

ACCA. (n.d.). Professional accountants – The future: 50 drivers of change in the public sector. Available at www.accaglobal.com/content/dam/ACCA_Global/Technical/Future/drivers-of- change-in-the-public-sector.pdf

Australian Government. (2019). *Tourism Forecasts*. Canberra, Australia Government.

Baume, M. (2019, August 8). Southbank tower to reach for the sky. Available at www.corelogic.com.au/news/southbank-tower-reach-sky (Accessed 2 April 2021).

City of Melbourne. (2016). *City of Melbourne Tourism Action Plan 2016–2019*. Melbourne: City of Melbourne.

City of Melbourne. (2019). Council plan 2017–2021: 2019 update. Available at https://mvga-prod-files.s3.ap-southeast-4.amazonaws.com/public/sitecollectiondocuments/council-plan-2017-21.pdf.

City of Melbourne. (2020). Southbank projects – City of Melbourne. Available at www.melbourne.vic.gov.au/building-and-development/shaping-the-city/city-projects/Pages/southbank-projects.aspx (Accessed 12 February 2020).

City of Melbourne. (2020a). Boyd community hub – City of Melbourne. Available at https://studiomimimoon.com/city-of-melbourne-boyd-community-hub-events/

City of Melbourne. (2020b). Southgate. Available at https://whatson.melbourne.vic.gov.au/search_2024?c=anything&dkw=any_day&date_range=&q=Southgate (Accessed February 14 (2020).

Crown Melbourne. (2020). *Crown Hotels, Accommodation, Casino*. Melbourne, Crown Melbourne.

DFO South Wharf. (n.d.). DFO South Wharf | Direct Factory Outlets. Available at www.dfo.com.au/south-wharf/ (Accessed 23 February 2020).

Herald Sun. (2017). Herald Sun | Breaking News from Melbourne and Victoria | Herald Sun. Available at www.heraldsun.com.au/ (Accessed 1 February 2020).

Maus, C. (2020). *Chapel Street Precinct Association Business Meeting*. Melbourne, Melbourne Chapel Street.

OECD (2013), OECD Framework for Statistics on the Distribution of Household Income, Consumption and Wealth. Paris: OECD. www.oecd.org/statistics/302013041e.pdf

OnlyMelbourne. (n.d.). Southgate Complex | Southbank. Available at www.onlymelbourne.com.au/southgate- complex-southbank (Accessed 14 February 2020).

Qpzm. (n.d.). Southbank demographics (VIC) local stats. Available at https://localstats.com.au/demographics/vic/melbourne/bayside/southbank (Accessed 13 February 2020).

Southbank News. (2020, February). Home | Southbank Local News. Available at www.southbanklocalnews.com.au/ (Accessed 12 February 2020).

Southbank Residents Association. (n.d.). History. Available at www.southbankresidents.org.au/history (Accessed 12 February 2020).

Southgate. (n.d.a). About us. Available at www.southgatemelbourne.com.au/about/ (Accessed 13 February 2022).

Southgate. (n.d.b). The history of Southgate. Available at www.southgatemelbourne.com.au/about/history (Accessed 12 February 2022).

Stanton, K. (2018, August 23). Southbank: What's to come for this flourishing inner-city neighbourhood. Available at www.domain.com.au/news/southbank-what8217s-to-come-for-this-flourishing-innercity-neighbourhood-20180821-h149jt-759572/

TravelVictoria. (2020). Southbank information – Travel Victoria: Accommodation & visitor guide. Available at www.travelvictoria.com.au/southbank/ (Accessed 11 February 2020).

VisitVictoria. (n.d.a). Malthouse theatre, attraction, Melbourne, Victoria, Australia. Available at www.visitmelbourne.com/regions/Melbourne/Things-to-do/Art-theatre-and-culture/Performing-arts/Malthouse-Theatre (Accessed 14 February 2020).

VisitVictoria. (n.d.b). Arts precinct, destinations, Melbourne, Victoria, Australia. Available at www.visitvictoria.com/Regions/Melbourne/Destinations/Arts-precinct (Accessed 14 February 2020).

Yarra River Business Association. (2019). *Yarra River Big Ideas Forum*. Southbank: YRBA.

3 The Possible Futures of Central Businesses District (CBD), Melbourne, Australia

Introduction

Melbourne is dynamic, innovative, creative, liveable, diverse and progressive (City of Melbourne, 2016b) and its identity is defined by the diversity of nationalities and cultures living and visiting the city. Melbourne's identity is also represented by a distinctive coffee and food culture that focuses on authentic experiences, and the city is famous for its creative atmosphere created by many national and international street artists and buskers, which is particularly noticeable in the lively nightlife and entertainment scene (City of Melbourne 2016b). Melbourne profits from a well-developed business culture which is represented by national and international micro, small, medium and large enterprises covering diverse industries such as the hotel industry and retail (City of Melbourne, 2020d).

Melbourne's Central Business District (CBD) attracts many national and international tourists and has currently the highest number of businesses and residents of any district in Melbourne (City of Melbourne, 2020d), but it is also confronted by multiple challenges and opportunities. This chapter investigates the dynamics between core CBD Melbourne and its micro- and small-sized enterprises (MSEs) which often lack the financial, social and political power of their larger competitors. However, MSEs are often the businesses that contribute to a city's identity and in return a city benefits from them. This chapter analyses the dynamics and developments occurring within the CBD in Melbourne that affect the possible futures of MSEs within its location. To achieve the aforementioned goal of the chapter the main research question is

> What are possible future scenarios for Melbourne's core CBD and its MSEs?

Diverse methods were used to investigate various aspects of the area and to answer the main research question. The current study included both primary (participant observation, interviews and newspaper analysis) data collection and secondary (desk) research. Qualitative data analysis was applied, which allowed the researchers to focus more deeply on people's emotions, stories, worldviews

DOI: 10.4324/9781003570240-4

and perceptions as opposed to dealing with concrete numbers, which present only one possible reality and do not show the external or internal circumstances or explain the context (Creswell & Creswell, 2017).Micro- and Small-Sized Enterprises in Core CBD

Most of the businesses and venues that contribute to Melbourne's identity are defined as MSEs. According to Gregory (2019), small enterprises are identified as for-profit enterprises that are independently owned and operated but that do not dominate their industry or local market. A subcategory, micro-enterprises, can also be described as the previous stage of a small enterprise because they may grow beyond their original foundation in the future. There is no universally exact definition with strict boundaries that limit a micro-enterprise or a small one. Given the opportunity and intention to grow, our criteria for MSE are not limited to the number of employees or the income they generate rather in terms of the impact they have on 'life in CBD' and their market.

The MSEs are clustered in categories and these categories are based on criteria related to the findings of the field research within the CBD area: *creative CBD, hospitality venues*, *tourist enterprises*, *retail*, *nightlife and entertainment*, *accommodation* and *social ambassadors*.

Life and Surroundings in Core Central Business District (CBD)

Melbourne has the reputation of being one of the safest and most liveable cities in the world, with CBD a dynamic commercial, entertainment and residential precinct representing Melbourne's business and financial centre (City of Melbourne, 2020d). According to the City of Melbourne (2020c), it encompasses the area from Flinders Street to Queen Victoria Market and from Spencer Street to Spring Street, plus the area between Victoria and Latrobe streets. The core CBD area in this research is defined by precinct boundaries and based on the researchers' findings. Furthermore, the described context of the CBD is based on the researchers' own experiences in terms of observation of life and surroundings in core CBD. Figure 3.1 shows the restricted area, from north to south, with the Flinders Street Station standing as a symbolic entrance into the CBD south precinct. According to the project manager of Rail Projects Victoria (RPV) (Personal Communication, 12 February 2020) the CBD south precinct is also known by Melbournians as the Flinders Quarter.

Notable precincts include the Greek precinct in the upper northeast area of the core CBD and the Chinatown precinct, both of which are located in the CBD north, the most populous area of the CBD.

According to the project manager of RPV, this is a popular residential area where many international students live. Initially, the gold rush in the 1850s brought settlers from both communities, the Greek and the Chinese, to Melbourne (Greek community, 2020; Live in Melbourne, 2020). Nowadays, Melbourne's Chinatown is one of the biggest Chinese communities in the Western world and is well-known for its Chinese Museum, its variety of Asian cuisine and its Chinese New Year celebrations with parades throughout the CBD (Live in Melbourne, 2020).

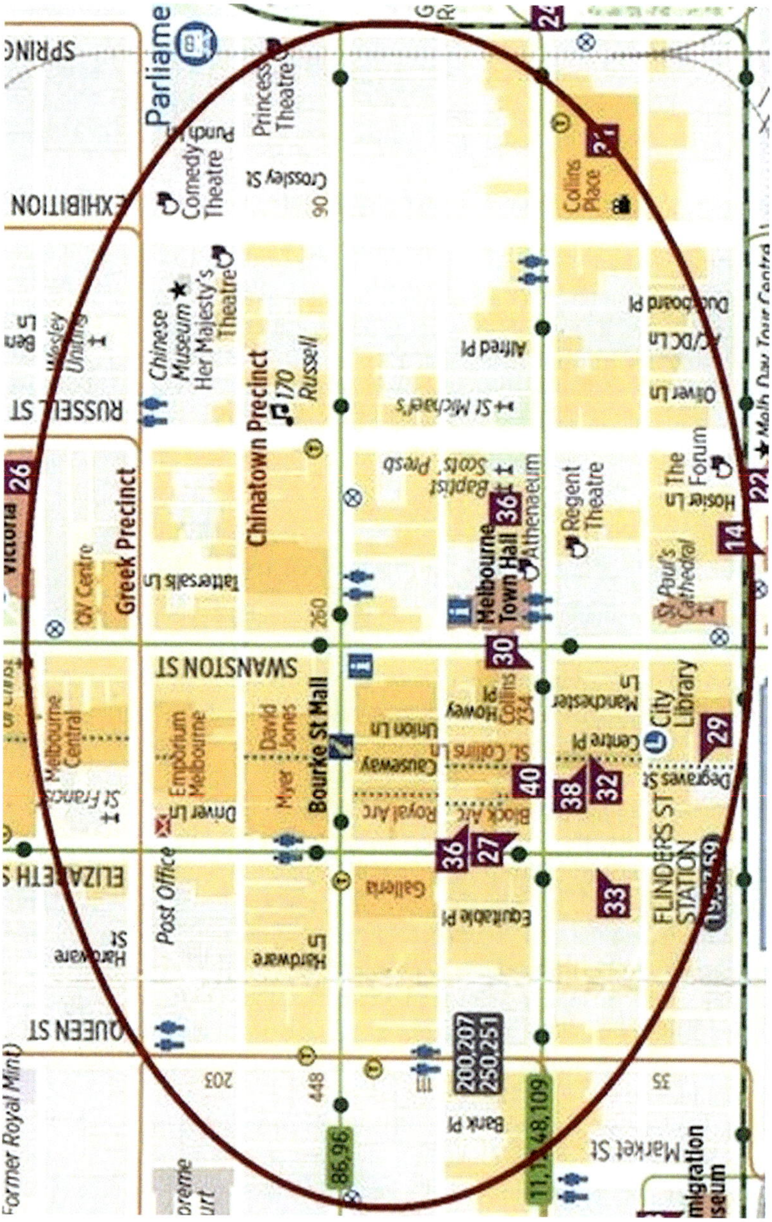

Figure 3.1 Map of core CBD.
Source: Visitmelbourne, 2020.

The core CBD operates 24 hours a day, serving a wide variety of residents, workers and visitors. The researchers' observations recorded real-life and real-time perspectives in Melbourne's CBD in terms of 'see, hear, smell and feel' the environment. Multiple strata contain businesses, residential apartments, narrow lanes and wide streets, along with squares containing important public and state operators that serve the city, such as the Town Hall and the Victoria State Library. Although CBD is densely built, the streets and lanes create a spacious environment surrounded by green areas, such as lawns attached to the library or universities, small parks and trees. To prevent too much road traffic within the CBD, Victoria State Government provided the 'free tram zone area' covering the core CBD completely (Victoria State Government, 2020a).

Outside working hours, CBD life continues with a large variety of restaurants, cafés and cultural entertainment opportunities open until late. Rooftop bars and clubs remain open until early morning with a wide variety of attractions that fit different interests in music, drinks, environment and atmosphere of people discovering CBD. The variety of people and the different facilities confirm the liveable dynamic image that Melbourne has created.

Key Drivers of Change

The CBD is currently undergoing enormous changes and development projects. Most of them have been examined during participant observation, interviews and meetings with companies and retrieved from desk research. A major development force for the city of Melbourne is the growing population and an increasing number of tourists. It is estimated that Melbourne will soon become the biggest city in Australia, with a predicted population growth from five million to six million people, and will replace Sydney in 2026 as Australia's most populated city (Mccrindle, 2020).

The CBD is defined as the business and financial centre of Melbourne (City of Melbourne, 2020d) and is characterised as dynamic and chaotic. The CBD will not only be a place for companies but also an area for different stakeholders, including residents and tourists, as Melbourne is internationally recognised as a creative city (City of Melbourne, 2018b) with its diverse range of street art, events, festivals, vibrant culture and nightlife (UNESCO, 2016).

Instead of being solely a business centre, CBD is also attracting increasing numbers of residents each year with the aim of making CBD the most rapidly growing small area in Australia (City of Melbourne, 2020e). Many different developments are visible because of the ongoing infrastructure developments. The availability of free Wi-Fi makes it attractive for residents and visitors (Victoria State Government, 2020b), and encourages employees and students to work outside – in parks, alleys and lanes. The expansion of the high-rise buildings and the Metro Tunnel Project are evidence of transport and residential improvements (State Government of Victoria, 2019), as is the building of two underground new metro stations in the CBD at each end of Swanston Street which will improve accessibility to the busiest and most vibrant area of the city. According to the project manager of RPV, large retail and food chains will be established in those metro stations. Some negative impacts are possible, such as access issues, noise pollution, parking difficulties and an increase in traffic congestion.

Climate change, in combination with population growth, is putting the resilience of Melbourne to test as all the changes happening in CBD are interrelated. Global warming puts pressure on destinations such as Melbourne, and the consequences are noticeable (State Government of Victoria, 2014; as the city has faced frequent bushfires, heatwaves, flooding and storms in recent years (Nicholls, 2020). According to *The Guardian* (2020), "7.7 million hectares in the south-eastern states, claimed 33 lives and almost 3000 homes" because of Australia's destructive fire season of 2019 and 2020. Possible consequences of climate change were identified from participant observation and interviews. As mentioned by the Minister for Trade, Tourism, and Investment (2020), "this is make or break for many businesses and tourist hot spots and not just in those areas directly hit by the bushfires." To support the victims and companies affected by the natural disasters, several different projects have been implemented, one example, mentioned by the director and manager of I'm Free Walking Tours, is Empty Esky – "a national movement of foodies and adventurers on a mission to support small businesses affected by the Australian fires" (Empty Esky, 2020). Sustainable solutions, such as the new metro stations, a bicycle plan and the free tram zone will help to decrease traffic and carbon emissions in core CBD (City of Melbourne, 2016a; State Government of Victoria, 2019). The *Climate Change Mitigation Strategy* and *City Switch* programme are important developments to help Melbourne to face global warming, showing that there is a growing awareness of climate change and sustainability (City of Melbourne, 2017; Clark, 2019). According to Wahlquist (2019), "Melbourne has become the first city in Australia to have all of its council-owned infrastructure powered by renewable energy. An example is the wide range of public transport, for example, it reduces greenhouse gases and resources such as oil."

Market Trends

A number of major trends are taking place in CBD. The first notable is globalisation, observed by participant observation and picture analysis performed by van der Zwan, Pétursdóttir and Werl (2020). According to the authors, "expansion of international chains is a symbol of globalization and consumerism," and technology can be seen as part of globalisation worldwide (Narula, 2014). In 2017, Melbourne was ranked 10th out of 98 cities listed in the 'Smart Cities Index' (Invest Victoria, 2019). Technology progress can be noticed everywhere, one example being that some traffic lights in Melbourne have lights embedded in the pavements to warn pedestrians distracted by their phones (ABC Radio Melbourne, 2017).

According to the general manager of the Chapel Street Precinct Association (CSPA, 2020; Interview 20th February 2020), a health and wellness trend is common in the CBD. "Where there is a developing trend towards a non-exclusive diet of healthy food of good quality, cooked simply" (Fine Food Australia, 2020). It was observed that CBD contains a variety of beauty and wellness salons and the notable presence of the food and coffee culture in core CBD. There is also an emergence of new opportunities in hospitality and retail such as rooftop bars. The 'Green Our Rooftop project' is a result of the Green Our City Strategic Action Plan by the City of Melbourne (2017) which is in line with their sustainability approach and is supported by the Victorian Government and Melbourne Water. Quoted by Urbanalyst (2015), "Rooftops in central Melbourne make up 880 hectares of space." This space is covered by solar panels, green spaces and rooftop bars, providing an opportunity for hospitality venues to grow.

Challenges and Opportunities of CBD

The main challenge is to meet the needs and expectations of diverse stakeholder groups. One of the challenges CBD currently facing is finding a balance between the growing local population, an increasing influx of tourists and the expansion of businesses within the city. Noise pollution and scarce residential spaces may cause problems in the future. The capacity of core CBD is limited and in the future the area may not be able to handle the rapid growth of public transport, residents, tourists, businesses and buildings without further regulation. Homelessness may rise in the future, for example, because of high rents (Henriques-Gomes, 2019) which could undermine the image of the core CBD.

Chinese outbound tourism has become the largest tourist segment in the world (Dichter et al., 2018) and according to Tourism Australia (2020), the Chinese are "the first largest inbound market for visitor arrivals and the largest market for total spend and visitor nights," with Chinese travellers creating 0.6 percent of the country's annual GDP (Cheng, 2020). The recently occurred coronavirus has affected the Australian tourism industry even more than the bushfires. The Australian Tourism Industry Council (2020) indicated that a one-month ban on incoming Chinese tourists caused more than $1 billion loss of Chinese tourism revenue, accompanied by booking cancellations and no-shows of both other international and domestic travellers across the regions. The current major issue – Coronavirus – has particularly affected Chinatown in CBD, as was mentioned in several participant observations and interviews, indicating that a reliance on a single market can be problematical.

The Metro Tunnel Project is an opportunity to decrease the pressure on the Free Tram Zone. According to the *CBD News*, the area will be even more accessible by, for example, the use of 'beacon technology.' Cited in the City of Melbourne (2019), "New beacons along Bourke and Swanston streets will send audio messages about potential obstacles to users' phones, to help people with low vision or blindness get around the city," a feature observed during the participant observation and newspaper analysis.

Stakeholders of MSEs in Core CBD

Melbourne's identity is typified by significant input from local communities, including residents, workers and business owners. To support small businesses, the

City of Melbourne has run the Small Business Grants Programme and financially supported approximately 380 small businesses to date (City of Melbourne, 2018a). The main reason for the city prioritising this programme was to boost the local economy by helping new operators to kick-start their business and support unique and innovative business models. The City of Melbourne sees itself as the mediator between smaller and bigger businesses and provides smaller enterprises with valuable business contacts and chances for corporations (City of Melbourne, 2018a). The City of Melbourne and other main stakeholders for MSEs in the core CBD are shown in Figure 3.2.

Figure 3.2 displays the main stakeholders in dark blue and additional stakeholders in light blue.

RPV creates opportunities for CBD Melbourne and its MSEs, and the new metro stations, scheduled for completion by 2025 (MetroTunnel, 2020), will improve accessibility, increase the number of potential customers for MSEs and partially solve traffic problems. The role and power of RPV are significant for the area and MSEs in directly shaping their possible futures. Additionally, during research it became obvious that the media plays a leading role in the formulation of the local image and provides communication between different stakeholder groups. Close collaboration was noticed between *CBD News* and the key businesses and individuals who influence the future of the CBD area. RPV has a dedicated column for each edition, utilising the paper as a tool to communicate main updates on the Metro Tunnel Project to the public and identifying the MSEs most vulnerable to the changes related to the construction works (Rail Projects Victoria, 2020).

It was noticeable that at least some major developments and changes in core CBD are directly connected to climate change. Some of the topics discussed during the participant observation and interviews included unpredictable weather and bushfires. Those factors not only impact the affected regions but also define the near future of businesses and areas highly dependent on tourists. For example, the smoke that covered Melbourne in January 2024 caused most constructions to cease operation due to the hazardous air quality (Hill, 2020), and MSEs around Melbourne are currently experiencing particular losses due to the decrease in tourism.

The Possible Futures

Different stakeholders, key drivers and trends influence the future of MSEs in Melbourne's CBD. Because of the different MSE categories and their specific requirements, the possible futures are set in three time frames, short term (0–2 years), medium term (2–5 years) and long term (5+ years).

Population growth and the increased influx of tourists in core CBD provide opportunities for MSEs to grow, although many MSEs are facing challenges that influence their possible futures. Rents for facilities are rising, which makes it harder for a micro or small business to afford a space. In addition, more people encounter social problems which may increase the number of homeless people in CBD in the medium term.

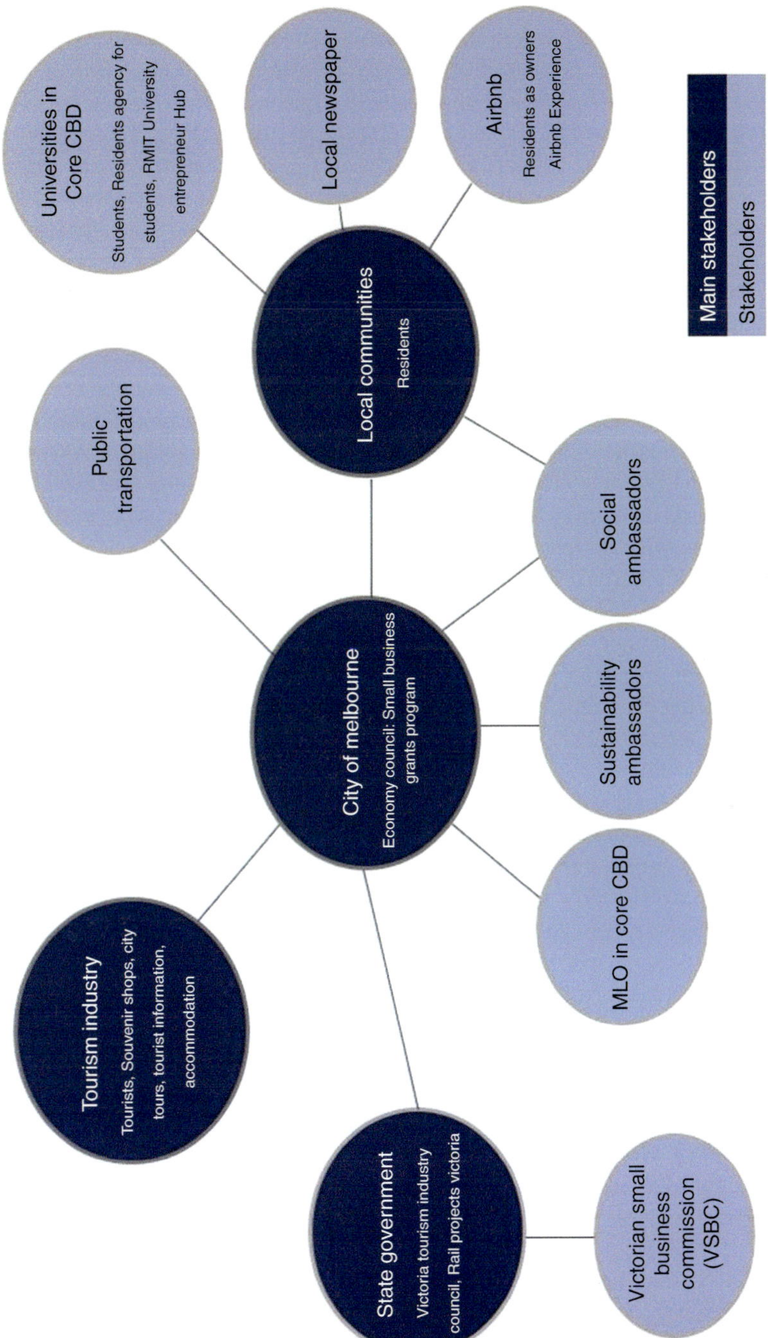

Figure 3.2 Stakeholder model of MSE in core CBD.

Source: Created by the authors.

An increase in Airbnb rentals, as part of *Accommodation*, can be expected in the next three years, because of the influx of tourists and consumers' interest in looking for a more unique and local experience. The friction that already exists may increase and affect the currently friendly atmosphere in core CBD, which would not be beneficial for MSEs doing business in the area. On a short-term basis, the construction going on in CBD makes it difficult to find MSEs and causes noise disruption in the area. The supportive character of the government towards MSEs in terms of compensation for the disruptions and grants (e.g. Small Business Grants Programme) is selective to some extent, and the possibility exists that some MSEs will not survive and will be eliminated.

Once the infrastructure developments are finalised in 2025, most disruption factors will disappear and the MSEs in core CBD will be more easily accessible. Competition for *Retail* and *Hospitality venues* MSEs will arise because larger organisations (MLO) will be opened on top of the new Metro Tunnel Stations, although because of Melbourne's strong identity in terms of consumers' interest in the specific characteristics of MSEs (atmospheric, sustainable, unique, innovative and creative), MSEs may survive this competition. Also, the high-rise buildings covering the CBD area provide considerable potential for more *Nightlife and entertainment* and *Hospitality venues* to be developed on rooftops.

Other possible futures relate to enhanced technology such as Wi-Fi and Beacon technology, which will ensure that everyone has equal opportunity to access MSEs in core CBD although technology might disadvantage other users. Currently, street-performers are under pressure due to technology changes enabling cashless payment reducing cash tips, but it might be possible to make wireless card reader devices more accessible for artists so that they can offer an alternative payment system to their audience.

Marketing strategies affected by climate change could result in an increase in sustainable practices (e.g. CitySwitch programme) in the long term. New features such as e-bikes will contribute to the sustainable approach and contribute to flexibility and accessibility to get around in CBD easily. Some MSEs, particularly *Touristic enterprises, Retail* and *Hospitality venues* in Chinatown, are at risk because of the Coronavirus pandemic and its restrictions as noted earlier, particularly those which focus on Chinese tourists and students as their main clients.

Conclusion

A universal approach to portray an overall possible future that covers all MSEs in the core CBD of Melbourne is difficult but not impossible. To overcome threats from emerging Medium and Large Tourism Organisations, regulation and support by the government are important tools. To gain such support initially, the main characteristics of the MSEs in line with what characterises Melbourne's identity need to be seen as the key to ensuring thriving businesses in this area. To stand out, creativity, innovation and uniqueness are vital components for MSEs to grow within the challenging dynamics of CBD. Key drivers of change are often interrelated, therefore, future scenarios are flexible and can be expected to change over time. Some MSEs will grow, and some will be eliminated, which in turn provide opportunities for new start-ups. Since Melbourne's image and characteristics are found rooted in many MSEs in Melbourne's CBD, the chance that consumers' current interests will remain with them in the future seems rather plausible.

References

ABC Radio Melbourne. (2017). Lights installed in Melbourne footpath to help distracted pedestrians cross safely. Available at www.abc.net.au/news/2017-03-27/lights-in-the-footpath-for-distracted-pedestrians/8391046 (Accessed 18 February 2021).

Bruno, A. Davis, K. & Staib, A. (2022) The Recovery in the Australian Tourism Industry. Available at www.rba.gov.au/publications/bulletin/2022/dec/the-recovery-in-the-australian-tourism-industry.html

Cheng, M. (2020, February 13). Australia needs Chinese visitors: Economic impact of coronavirus travel ban could be huge. Available at www.scmp.com/week-asia/opinion/article/3050243/australia-needs-chinese-visitor s-economic-impact-coronavirus (Accessed 22 February 2020).

City of Melbourne. (2016a). City of Melbourne Tourism Action Plan 2016–2019. Available at www.academia.edu/36248173/CITY_OF_MELBOURNE_TOURISM_ACTION_PLAN?auto=download

City of Melbourne. (2016b). Future Melbourne 2026 Plan. Available at https://participate.melbourne.vic.gov.au/future

City of Melbourne. (2017). *Green Our City Strategic Action Plan 2017–2021*. Melbourne: City of Melbourne. Available at https://mvga-prod-files.s3.ap-southeast-4.amazonaws.com/public/sitecollectiondocuments/green-our-city-action-plan-2018.pdfCity of Melbourne. (2018a). Another boost for Melbourne's small businesses and start-ups. Available at www.melbourne.vic.gov.au/small-business-grants

City of Melbourne. (2018b). City of Melbourne creative strategy. Available at https://s3.ap-southeast-2.amazonaws.com/hdp.au.prod.app.com-participate.files/7315/3077/4837/City_of_Melbourne_draft_Creative_Strategy.pdf (Accessed 23 February 2020).

City of Melbourne. (2019, December 2). Beacons boost city access for people with low vision. Available at www.miragenews.com/beacons-boost-city-access-for-people-with-low-vision/

City of Melbourne. (2020c). Development activity model. Available at www.developmentactivity.melbourne.vic.gov.au (Accessed 14 February 2020).

City of Melbourne. (2020d). Melbourne. Available at www.melbourne.vic.gov.au/

City of Melbourne. (2020e). Melbourne facts and figures. Available at www.melbourne.vic.gov.au/data-and-insights

Clark, G. (2019). Melbourne becomes first city with all council infrastructure powered by renewables. Available at www.theguardian.com/australia-news/2019/jan/17/melbourne-becomes-first-city-with-all-council-infrastructure-powered-by-renewables (Accessed 17 January 2020).

Creswell, J. W., & Creswell, J. D. (2017). *Research Design: Qualitative, Quantitative, and Mixed Methods Approaches*. California, Sage Publications.

Dichter, A., Chen, G., Saxon, S., Yu, J., & Suo, P. (2018, September). Chinese tourists: Dispelling the myths. Available at www.mckinsey.com/~/media/mckinsey/industries/travel%20logistics%20and%20infrastructure/our%20insights/huanying%20to%20the%20new%20chinese%20traveler/chinese-tourists-dispelling-the-myths.pdf (Accessed March 2022).

Empty Esky. (2020). About Empty Esky. Available at www.emptyesky.com.au/about/ (Accessed 15 February 2020).

Fine Food Australia. (2020). Let's talk Melbourne food trends. Available at https://finefoodaustralia.com.au/industry-updates/lets-talk-melbourne-food-trends/ (Accessed 19 February 2020).

Greek community. (2020). A brief outline of the history of the Greek orthodox community of Melbourne Victoria. Available at www.greekcommunity.com.au/gocmv_public/index.php/en/about-us (Accessed 20 February 2020).

Gregory, A. (2019, June 4). Differences between micro business and small business. Available at www.thebalancesmb.com/differences-between-micro-business-and-small-business-4159188 (Accessed 8 February 2020).

Henriques-Gomes. (2019). Homelessness becoming concentrated in Sydney and Melbourne, study finds. Available at www.theguardian.com/australia-news/2019/may/30/homelessness-becoming-concentrated-in-sydney-and-melbourne-study-finds (Accessed 30 May 2020).

Hill, M. (2020, February). Smoke shuts down city construction. *CBD News*. Available at www.cbdnews.com.au/smoke-shuts-down-city-construction/ (Accessed 3 March 2020).

Invest Victoria. (2019). Melbourne Tech City. Available at www.invest.vic.gov.au/news-and-events/news/2019/feb/melbourne-confirmed-as-australias-tech-city

Live in Melbourne. (2020). Melbourne's Chinese community. Available at https://liveinmelbourne.vic.gov.au/discover/multicultural-communities/chinese (Accessed 20 February 2020).

Mccrindle. (2020). Melbourne reaches 5 million. Available at https://mccrindle.com.au/article/topic/demographics/melbourne-reaches-5-million/

MetroTunnel. (2020, January 25). New underground stations. Available at https://metrotunnel.vic.gov.au/about-the-project/stations (Accessed 20 February 2020).

Narula, R. (2014). *Globalization and Technology: Interdependence, Innovation Systems and Industrial Policy*. Cambridge, John Wiley & Sons.

Nicholls, N. (2020, February 10). Fires and floods: Australia already seesaws between climate extremes – and there's more to come. Available at www.theguardian.com/commentisfree/2020/feb/10/fires-and-floods-australia-already-seesaws-between-climate-extremes-and-theres-more-to-come (Accessed 18 February 2020).

Rail Projects Victoria. (2020, December 10). Our projects. Available at https://railprojects.vic.gov.au (Accessed 22 February 2020).

State Government of Victoria. (2014). The vision for Melbourne: Melbourne will be a global city of opportunity and choice. Available at www.planning.vic.gov.au/guides-and-resources/strategies-and-initiatives/plan-melbourne (Accessed 13 February 2020).

State Government of Victoria. (2019, October 17). Construction in Melbourne CBD. Available at https://metrotunnel.vic.gov.au/construction/cbd (Accessed 14 February 2020).

The Guardian. (2020). Inside Australia's climate emergency: The new fire zone. Available at www.theguardian.com/environment/ng-interactive/2020/feb/12/living-in-the-climate-emergency-australias-new-fire-zone (Accessed 20 February 2020).

Tourism Australia. (2020). China. Available at www.tourism.australia.com/en/markets-and-stats/market-regions/greater-china/china.html (Accessed 14 February 2020).

UNESCO. (2016). UNESCO creative cities network. Available at https://en.unesco.org/creative-cities/melbourne (Accessed 18 February 2020).

Van der Zwan, R., Pétursdóttir, R., & Werl, J. (2020). *What Is the Current Situation of the Tourism Industry within the Inner CBD in Melbourne and What Are Plausible Future*

Scenarios for MLO's in the Touristic Context? Breda: Breda University of Applied Sciences.
Victoria State Government. (2020a). Melbourne's free tram zone. Available at https://conference.acen.edu.au/2020-free-tram-zone/ (Accessed 18 February 2020).
Victoria State Government. (2020b). Victoria's free public wi-fi network. Available at www.vic.gov.au/victorias-free-public-wifi (Accessed 18 February 2020).
Visitmelbourne. (2020). Travel information. Available at www.visitmelbourne.com/
Wahlquist, C. (2019, January 17). Melbourne becomes first city with all council infrastructure powered by renewables. Available at www.theguardian.com/australia-news/2019/jan/17/melbourne-becomes-first-city- with-all-council-infrastructure-powered-by-renewables (Accessed 25 February 2020).

4 The Possible Futures of Dockland, Melbourne, Australia

Introduction

In this chapter, the focus is on Docklands, one of the precincts most affected by COVID-19 and which was already experiencing several challenges prior to the pandemic. The Docklands development could be described as a mega-project. Renovation and transformation of a declining water port into a modern waterfront are not unique to Melbourne and have occurred in other cities around the world. The basic aim is always the same: to compete on the world stage and attract investments, so, cities need to create a new spectacular area with a strong image and marketing identity in order to make a transition from industrial to 'post-industrial' economy (Shaw, 2013). However, according to comments in the literature and diverse local newspapers, and talking with residents, Docklands is generally described as not reaching its potential, and arguments were expressed that other changes would have to happen if Docklands wants to become a more attractive place. Accordingly, the aim here is to understand and reveal the possible futures of Docklands, focusing on the strategic position of its medium and large organisations (MLOs) as well as residents.

Context

Demographic

The population of the Docklands district is 10,964 (in 2016) of whom 51 percent are male and 48.5 percent female. The median age is 30 years, which is considerably lower than the national (38 years) and regional (Victoria) average (37 years). Most people (52.2 percent) are aged between 20 and 35 years, which is substantially larger than the national (22 percent) and regional (21.1 percent) average, meaning that the Docklands area is an area with young residents. There are 2,556 families registered, and families with children have an average of 1.3 children. The median weekly household income is AUS$1,868 (national weekly income AUS$ 1,924.60) with a median monthly mortgage payment of AUS$2.167, and a median weekly rent of AUS$501 (Australian Bureau of Statistics, 2017).

DOI: 10.4324/9781003570240-5

Environmental

The Docklands area is bordered by the Yarra River in the south and the Moonee Ponds Creek in the west and is located west of the Central Business District (CBD) of Melbourne, separated by railways and state route 55. In the centre of Docklands is Victoria Harbour, a yacht marina, surrounded by promenades, restaurants, high-rise residential and commercial buildings, including office and retail developments. The north part of Docklands is known as an entertainment area and the south hosts many large companies (Travel Victoria, n.d.). In total the area is 190 ha which includes 44.7 ha of water and 7 km of waterfront. It is made up of eight diverse precincts (Travel Victoria, n.d.).

The *Waterfront City* precinct (orange) is the retail and entertainment area, with the Southern Star Ferris wheel and a shopping centre. It also offers many dining areas and residential homes. *New Quay* (yellow) is a residential and commercial area on the waterfront. It is considered a high-income area with gourmet dining experiences, modern buildings and contemporary art features. The *Digital Harbour*

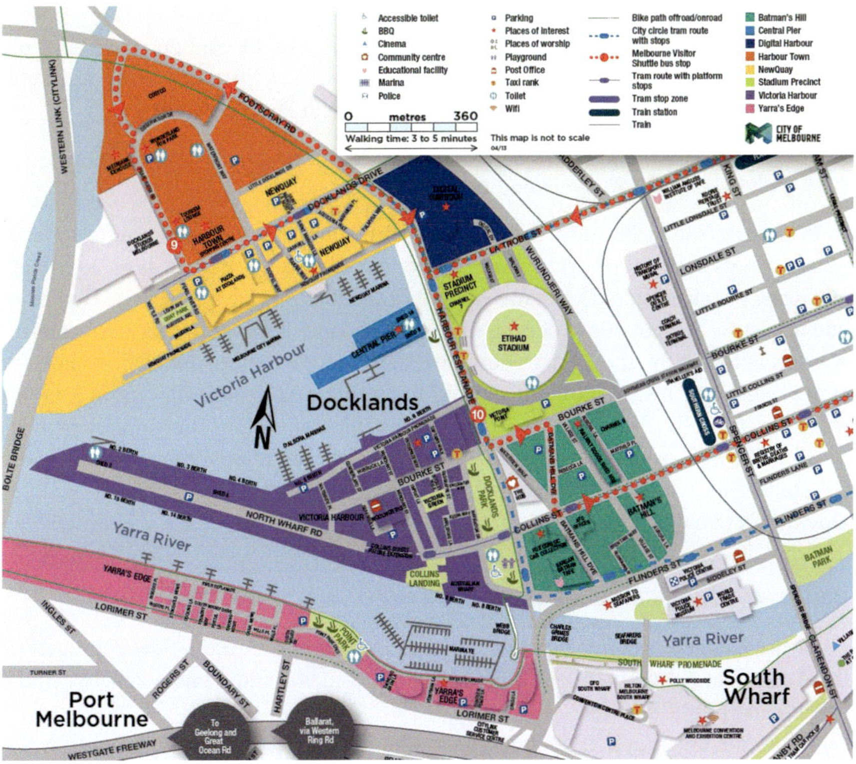

Figure 4.1 Visualisation of different precincts.

Source: Maps Melbourne.

(dark blue) precinct houses various tech companies, residing in modern designer buildings. The *Stadium* (light green) is a dominant feature of Docklands, and hosts sporting events and concerts. It is easily accessible due to the railway station on the east and the pedestrian bridge over the railway connecting the central business district and the Docklands. The *Central Pier* (light blue) is the centre of the Docklands with a luxury super yacht marina and luxury wine and dining experiences. The *Victoria Harbour* precinct (purple) is in the south of the Victoria Harbour and mostly consists of residential commercial buildings. The Docklands Park is easily accessible from the Victoria Harbour precinct and is family- and children-friendly. The *Batman's Hill* (turquoise) precinct is also a combination of residential and retail buildings. Melbourne's history is visible in this precinct, as it was the starting point of development in Melbourne. Finally, the *Yarra's Edge* (pink) precinct is described as progressive and inspirational by Family Getaways Melbourne. It has a large promenade with restaurants, cafés, art galleries and shopping opportunities. The Webb Bridge connects pedestrians from this precinct with the rest of the Docklands across the Yarra River (Travel Victoria, n.d.).

Phases of Developments

The Docklands have experienced many developments throughout the last three decades. According to Shaw (2013), Docklands had key periods during its development, between 1990 and 1999, 2000 and 2005 and 2010 onwards.

1990–1999: Economic Situation and Jeff Kennett

In the years from 1990 until 1999, Docklands attracted attention due to an economic revitalisation of Melbourne. The development project was a private business, which attracted large investments and other private businesses. However, due to a slow start to construction, investors became less interested, until interest was recaptured by the construction of the stadium, street extension and the approvement of a development plan to construct and design the world's tallest structure. Initially, then Premier Kennet was against investing state money in these projects, however, he was pressured to support and organise funds to invest in the developments which led to international investors becoming interested again in Docklands (Shaw, 2013).

2000–2005: Social, Environmentally Sustainable Impacts

After the Kennett premiership ended, Steve Bracks introduced a new approach by the Australian government, with the focus on residents and a balance between the private and public sectors.

While the construction contracts from the Kennett era were still operating and building continuing, Bracks also introduced sustainable regulations involving international contractors such as Lend Lease and ING Real Estate Development, with expertise in social and economic developments (Shaw, 2013).

By 2003, approximately 3,000 living spaces had been built. However, it was estimated by Lend Lease that potential residents would have to have about twice the average household income to be able to rent the newly developed spaces in Docklands. During this period, because of the large number of living spaces being built, the market for apartments was contracting. New initiatives were taken to provide subsidies for social housing, enabling lower-income residents to rent out these spaces (Shaw, 2013).

However, the implementation of government funding was not immediate, which led to investors retreating from Docklands by 2004 (Shaw, 2013). Consequently, the delay in funding and the high prices of spaces resulted in negative consequences for the area's economy, and the environmental plans were halted.

2010–Present Day

The City of Melbourne has created multiple themes and criteria by which to develop Docklands, which would benefit the stakeholders involved. In October 2010, the *Shared Vision for Docklands* was launched for the period until 2020. The themes were: 'Theme 1: Capturing the Essence of Melbourne,' 'Theme 2: Embracing Melbourne's waterfront' and 'Theme 3: Creating a 21st-century city' (City of Melbourne, 2012). In addition to these themes, several specific goals were set for each theme. These 'goals' were set up and supported by the community and experts.

The City of Melbourne (2021) developed a 'Docklands Community and Place Plan' which lists all the ideas and future plans for Docklands.

Currently, according to the development authority, on completion (in 2025) the Docklands will have absorbed AUS$ 17.5 billion of private investment. Just before COVID-19, 14.6 billion AUD of private investment had been made and 73,000 people were working in the area (Victoria Development, 2019), mostly in insurance and finance, commerce and public administration (Shaw, 2015). The construction of the Docklands, which is still under completion, has employed about 3,000 construction workers to work on the project annually (Victoria Development, 2019), with some of the largest Australian companies located in this area.

Currently, Docklands has 36 works of art in its parks, promenades and architecture, and the cultural diversity can be seen everywhere in murals (Curran, 2018). It is like an open-air outdoor art gallery, with contemporary art sculptures widespread throughout the Docklands. The sculptures have become a major feature in the area representing the creative heart of the Docklands and indeed of Melbourne (Lim, 2020).

Research Approach

Definition of Stakeholders

MLOs located in Docklands are organisations that have at least 50 employees, and identify themselves as medium or large, and are located in Docklands. Their annual budget should be above €10 million and/or the number of employees should be above 50. Large organisations can be multinational corporations or governmental agencies, they are likely owned by anonymous shareholders, therefore ownership is less emotionally involved than for the 'owner-manager' category and is more likely to be bureaucratic with large financial influence (Schaefer, 2010).

Residents are persons residing in Docklands permanently or continuously, and that population includes everyone who lives and has lived in the past year or who works and has worked in the past year in the Docklands area. As it is not always possible to determine the specific place of residence using online research methods, all persons defining themselves as residents were included in data collection.

MLOs and residents were contacted and individuals and companies messaged with posts on community Facebook pages connecting to expats living in Docklands. Guest lectures given by industry contacts also provided much insight into Docklands and MLOs, and secondary data from local newspapers and other written interviews were used to create a thematic analysis. Data were used in the identification of common themes, which were added to the key drivers of change.

Analysing the Data

Thematic analysis has become a popular qualitative data analysis method according to Braun and Clarke (2012), who explain that thematic analysis is useful to analyse themes or patterns that might have relationships with the data collected. Additionally, thematic analysis is effective in discovering and studying frequently used themes in the comments of the segment, and has the ability to compare relationship between various themes.

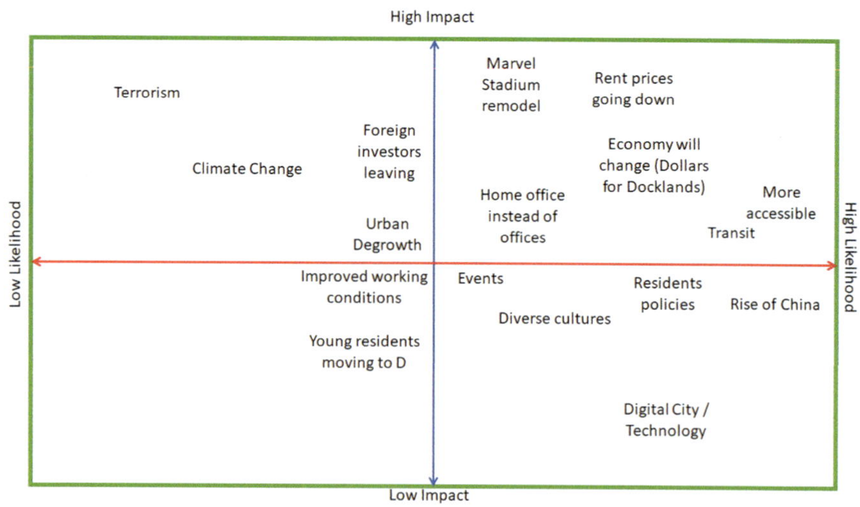

Figure 4.2 Themes.

Source: Created by the authors.

Figure 4.2 shows the results and items (themes) that were observed and discovered from the primary and secondary data, and the data are sorted by the impact and likelihood of the topic.

In the next section the four main themes which will influence the MLOs and residents are given detailed attention. The first concerns the 'Dollars for Docklands' initiative intended to grow the economy in Docklands, especially encouraging tourism by handing out vouchers to tourists staying in Docklands. The second theme is 'connectivity within Docklands' and other Melbourne precincts reflecting the environmental barriers to travel. The third theme is focused on the 'Marvel Stadium,' the different opinions about it, and its development potential, reflecting that it is a dominant structure in Docklands. The fourth theme deals with the changed image of Docklands, given that COVID-19 has changed its image and perception, with one of the most notable impacts of COVID-19 being working from home. This has affected many MLOs and residents, with numbers leaving the area facing increased rent prices.

Theme 1: 'Dollars for Docklands' Initiative to Grow the Economy in Docklands

During the pandemic, many businesses struggled to keep their doors open. This is especially true with tourism-related businesses, because Australia closed its borders to prevent the spread of the virus. The government created many initiatives to help support these businesses, one of which was the 'Job Keeper' payment. Job Keeper became available on 28 September 2020 and lasted until 28 March 2021.

Coincidentally, the Docklands Chamber of Commerce (DCC) introduced the 'Dollars for Docklands' which was launched at the end of March 2021 (Car, 2022). This was an initiative introduced by the DCC involving organisations in Docklands in promoting and supporting struggling local businesses and accommodations, by implementing a voucher/money back system. Many stakeholders became involved, particularly in the accommodation sector, the service-providers and entertainment sector.

The Dollars for Docklands was not widely known nor discussed in the media or news. Most of the reports found on the topic were found in www.docklandsnews.com.au along with some instances of locals discussing this initiative on social media.

The targeted stakeholders were mainly in the accommodations and service-providers sectors. Organisations were required to sign up on a dedicated website 'docklandsdollars.com.au' to take part in the initiative, under which domestic travellers (target group) can turn in vouchers they had been given. A 2-day accommodation was the minimum length of stay for visitors to become eligible for redemption vouchers worth up to $210.00 to be used in the Docklands area. The vouchers could be used for accommodation and other food and beverage purchases from the participating members (expert at DCC, 2021).

However, reports from residents online suggested some were not in favour of the Dollars for Docklands. Many were concerned about the impact that short-stay tourists might have on their way of life, "there is nothing in this scheme that will allow me to gain any benefit" (local male, 2021), "Airbnb issues, that create wear and tear on our premises" (local male, 2021), "I quite enjoyed during lockdown not finding cigarette butts or random objects dropped into my courtyard but now the return of short stayers has seen the return of inconsiderate people" (local female, 2021). These quotes conflict with the DCC, which stated that they have received support from both the businesses and the community, as "The residents don't want to see their business community die… this would impact their lives if the services were no longer available in Docklands" (expert at DCC, 2021).

Theme 2: Connectivity and Accessibility within Docklands and Other Precincts

The Docklands area has struggled in the past years with accessibility to the precinct. This can be attributed to the environment of the area. When looking at Figure 4.3, one can see that the precinct is surrounded by Highway M1/M2 (west and south) and the state route 55 (east and north) and additionally by the Yarra River in the south, the railway in the east and north, as well as the Moonee Ponds Creek in the west. Some of the residents and professionals commented on accessibility, although some residents perceived this as positive, mentioning that: "Docklands somewhat feels separate from the busy life of CBD, which is a plus in my personal opinion" (resident from Docklands).

The Yarra River

The Yarra River also divides Docklands itself. River cruises have been deemed outdated, and according to the senior officer of the Yarra River Business Association (YRBA), have become a thing of the past, since they were unlikely to survive the pandemic, and self-hire boats are becoming more popular. The Yarra River lacks a proper authority, and "there are many different authorities, and many places of legislation, that control out waterway," according to the senior officer of the YRBA, 2021. This means that it is highly complicated to build bridges and tunnels, and also to establish ferries linking Docklands. This feeling was also supported by Lord Mayor Robert Doyle in 2017 (Scanlan, 2017). In 2013 a plan was made to create commuter ferries across the Yarra River for workers and also for large events taking place in the Marvel Stadium (Places Victoria & City of Melbourne, 2013). This plan was strongly supported by different organisations, and the Victorian Transport Association chief executive Peter Anderson found that "using the Yarra as a means of moving people is a tremendous way to better access the city" (Scanlan, 2017). This commuter ferry service was not developed further because of the bureaucracy involved and confusion in governance.

Figure 4.3 Movement barriers around Docklands.
Source: Places Victoria & City of Melbourne, 2013.

Bicycle Lanes and Pedestrian Areas

Within Docklands the travel pattern has "very high proportions of walking, cycling and public transport use and low levels of car dependency" (Places Victoria & City of Melbourne, 2013). The number of residents and workers in the area have increased to 15,495 (Australian Bureau of Statistics, 2021). A short-term action from the City of Melbourne (2022–2023) has improved the pedestrian and cycling connections, as have an underpass and five other actions to improve the cycling and pedestrian experience in Docklands, including improved bike lanes at Adderley Street, Neil Street, Victoria Street, Collins Street in Docklands and Bourke Street. Cyclists as a proportion of vehicles entering the city grew from 13.3 to 14.7 percent in the 2014. A senior officer of Rail Projects Victoria (2021) claims that the bike lane infrastructure is being developed further.

> They are spending millions on it because many people would prefer to cycle in, than catch public transport. Before Covid 15% of all commuters rode their bikes into the city and it was growing, and it is probably even greater now.

This is verified by a senior officer of the YRBA (Personal Communication, 2021), claiming that "the pedestrian and cyclist management has been an ongoing problem… [cyclist] creating potential damage to pedestrians, and possibly fatalities." This problem of cyclists and pedestrians sharing communal space has initiated elaborate plans for bicycle users. One is to create 'cycling corridors' connecting different precincts, and from the Docklands to the CBD, and also to the west across the M1 and the Moonee Ponds Creek, with a 'cycling corridor' through the south of Docklands, connecting to South Warf and Fishermen's Bend (City of Melbourne, 2016).

Public Transportation

Accessibility was mentioned as the reason why new businesses are relocating to Docklands. "I think there are more people or more businesses moving to Docklands due to accessibility, public transport for workers and I think it's proximity to the city – the traditional CBD" (sales manager of tourist attraction, 2019). The public transportation network is being extended regularly, the metro tunnel is being built (2025) in order to ease pressure on routes, which are already at their maximum capacity and to connect Docklands with the CBD, South Wharf and South Bank. The city loop connects the east of Docklands with the rest of Melbourne at the "Southern Cross" station (senior officer of Rail Projects Victoria, 2021).

The senior officer of the Rail Projects Victoria claims that especially "the surface rail line very much divides the city at that location [the east]." An extension of the tram line from Collins Street over the Yarra River to Turner Street developed (City of Melbourne, 2022–2023). The Shopfront Activation Program, which concluded in March 2023, transformed 67 vacant shops into new retail spaces and enterprises in the CBD, Lygon Street and Docklands precincts. A total of 30 activations were completed through the City Activation Grants programme and that contributed to a return of the night-time economy that exceeded pre-pandemic levels. In addition, the City of Melbourne supported increased visitation to Docklands through related initiatives to strengthen economic recovery, tourism and events. One of the ongoing projects is delivering and managing the Business Precinct Program – investment of up to $90,000 to DCC for business support and activation initiatives (City of Melbourne, 2022–2023).

A new metro route from Mernda through the CBD, underneath the Yarra River to Fisherman's Bend and Newport will likely be established by 2030, which would supposedly eliminate the extension of the tram line from Collins Street. This proposed new metro line also garnered positive feedback as opposed to the tram bridge according to Docklands News and the Chairman of the Yarra Residents Action Group Keith Sutherland (Hill, 2020).

The lack in connectivity hinders employees to easily travel to Docklands, and often causes people to use a vehicle instead, which requires parking and other infrastructure. This impacts not only residents in the way they commute to work outside of Docklands but also employees working for MLOs in Docklands. These connectivity issues are being resolved, some faster than others, in order to avoid the need for extra parking, wider streets and more asphalt.

Theme 3: The Marvel Stadium

The Marvel Stadium, which has a capacity of hosting 50,000 people, was opened in 2000 and hosts sporting events, such as Australian football, cricket, soccer and concerts. This stadium is the largest infrastructure located in the Docklands and has contributed greatly to its development (Young Tourism Network).

> There has been a lot of time and effort going into making the Docklands nicer to be around. When it was first built it was like, just tall buildings. It wasn't much, to really go there for. And now they got like the big sports stadium, and they have gradually built up the rest.
>
> (Investor and hospitality entrepreneur in Dockland)

Besides being the largest infrastructure of the Docklands, the stadium has a huge importance for the whole city given how important sport is in Australian culture. This was confirmed by one resident of Melbourne talking about "general life in Melbourne" "To top it off, Sports. Melbourne is a very Sports crazy city. Australia as a whole is sports crazy. But especially Melbourne, the variety that you are getting from sports" (resident of Melbourne).

However, many argue that the stadium is wrongly positioned and therefore generates a many problems. The main one is that its location cuts the Docklands off from the CBD. This location has other repercussions, notably for the residents, as highlighted by an investor and entrepreneur in hospitality and catering: "we put a very big stadium right in the middle of it, so when we have football games there or concerts and they get 50,000 people it's like, it's max up all, if you are living there for instance, you don't want those people to be there."

In the last year on several occasions, different actors have been talking about the demolition of the stadium. In a forum held in 2015 talking about how Melbourne could look in 2035, the former Lord Mayor of Melbourne, Cr Robert Doyle, said that the stadium will be destroyed within 20 years (Masanauskas, 2015). He also proposed to rebuild a new stadium in Western Park which is also located in the Dockland area. According to him, this location would be more convenient compared with the current one as it won't cut off the Docklands from the CBD (Masanauskas, 2015). However, opponents argue that it's not the stadium that cuts off the two mentioned areas but the surface railway. The best solution would be to build over the railyards as that would connect the two areas better and give more space around the stadium. According to an investor and entrepreneur in hospitality this solution is realistic in a near future

> the other thing we have between the CBD and Docklands is railways lines, so what they are slowly doing is building on the top of the railways lines, so soon you won't see the railways lines, they are almost underground and they will be building on top… .

Collingwood (Australian rules) Football Club president Eddie McGuire also argues that the stadium should be demolished but instead of rebuilding a new stadium in the Docklands area, he proposes to do so on the site of the Hisense Arena (where the tennis stadium and cricket stadium are in the east of the CBD). The current location of the stadium in Collingwood would then be sold for commercial and residential development (as the stadium is owned by the Australian football league). This idea attracted the attention of Daniel Andrews, Victorian premier, who is willing to develop this area of the city further. This option was mentioned by the investor and entrepreneur in hospitality and catering as he said

> This talk that they are going to demolish the stadium, the stadium has only been there for 20 years but it's now been bought by the AFL, they own it, and they told me about possibly demolishing it the next 5 to 10 years, because it had more value to build more apartments, so they will built to stadium further out of Melbourne… .

It should be noted that in 2018 the Victorian State provided 225 million AUD to renovate the stadium (Schout, 2021). The idea will be to transform "a stadium which just held one game a week to an everyday community meeting place." A town square will be created inside the stadium which will be able to host several events such as a night market and other community gatherings. This square will be located in the part of the stadium facing the CBD. The idea of this transformation will be to make the stadium a place where people can gather and enjoy Melbourne life, thereby making the Dockland area more attractive (Schout, 2021).

Theme 4: Dockland's Changes Due to the Pandemic Work Behaviour Changes

According to institutes that ranked countries around the world by how well or badly they handled the COVID pandemic, Australia ranked in one of the top positions, for example, the Lowy Institute ranked Australia 8 out of 98 countries (Galloway, 2021). However, the city of Melbourne was hit hard by the COVID restrictions (Mark Doman, 2020), undergoing several lockdowns from hard ones to light ones, including a curfew in August 2024. Those restrictions affected particularly Melbourne as a cultural city and an event hub (Silva, 2020). As of December 2020, 52,000 peoples had lost their jobs in the city, which is 1 out of 10 working people (Capp, 2020). According to a consultancy firm, central Melbourne could lose up to AUD 110 billion in the gross local product over the next five years.

COVID-19 restrictions caused a major change to work arrangements for Victorian workers, with 20 percent of respondents working from home before COVID-19 growing to 45 percent working from home during the pandemic (The University of Sydney, 2020).

> The evidence reinforces the fact that as we move through and beyond the COVID-19 period, we can expect commuting activity to decline by an average of 25 to 30 percent as both employers and employees see value in a work from home plan.
>
> (Professor David Hensher, Founding Director of ITLS, The University of Sydney, 2020)

Throughout COVID-19, employees living in metropolitan areas worked from home more than their counterparts in regional areas, according to the survey (The University of Sydney, 2020). Starting from late April 2020 the state of Victoria started to lose more jobs due to the COVID-19 than any other Australian state (Borland, 2020). At the end of January 2021 when people were encouraged to come back to work in their office the occupancy rate of the Melbourne CBD was only 31 percent compared with that in Sydney of 45 percent (Morgan, 2020; Schlesinger, 2021).

Despite the measures ending, many companies seem to be in no hurry to have their employees return to the office and plan to keep flexible work arrangements into 2021. "There's still a big mountain to climb in terms of people coming back," said Jane Fitzgerald, executive director of the Property Council of Australia's NSW unit (World Economic Forum, 2021).

> There is no doubt the transition to working from home has been difficult, but it has also come with benefits such as no two hour per day commute which is more time with the family which has increased my well-being.
>
> (Professional Scientists Australia, 2020)

> With approximately 100,000 workers coming into Docklands pre COVID-19 underpinning our precinct, the lack of return workers is having a dramatic impact. Whilst 75% can return to the office currently, we are nowhere near that level. 20–30% are returning, however only for 1–2 days per week. This will continue to negatively impact Docklands. Workers have now been working from home for 12 months and have successfully managed their work/life balance and are not keen to return. Majority of Workplaces are not forcing workers to return and will continue on the flexible path for many months and in some cases permanently.
>
> (Docklands Chamber of Commerce president, 2021)

Everything has been changed by COVID-19, and DCC president believes that 50 percent of local businesses will close permanently. This would effectively eliminate the need for large office buildings. Docklands, especially the southern part which is filled with high-rise office buildings, many currently unused or underused, with high overhead costs, could suffer greatly (Dan, 2021).

In December 2021, half of the shopfronts in Docklands were closed or vacant making the area the worst affected by the COVID crisis in Melbourne (Rachel Eddie, 2021). As of 1 March 2021, the Dockland area was empty compared with pre-COVID times (Tuffield, 2021). Furthermore, business owners in the Dockland are worried that the area will never recover fully from the COVID crisis. Johanna Maxwell, DCC president, described today's situation as 'completely flat' and she doesn't see any change coming in a near future (Tuffield, 2021). She points out that it is not only the hospitality sector that is hurting but all other types of business (Tuffield, 2021). Even with the lifting of some restrictions recently, Maxwell is pessimist about people coming back to offices in number (Tuffield, 2021). In fact, in February 2021 even with restrictions being lifted, the number of people commuting to the Docklands was closer to those during the lockdown than those of 2019 (pre-COVID) (Rachel Eddie, 2021). However, perhaps people will remember that the entertainment and other assets of the Docklands are still present and therefore will start slowly to come back again in the area (Tuffield, 2021).

The idea of working from home is not possible for all workers and is not feasible in some industries. For example, only 12 percent of workers in retail, 13 percent in agriculture, 15 percent in transport and storage and 16 percent in manufacturing have the opportunity to work from home, and there are other industries that require a suitable workplace, and increased flexibility requires the right natural conditions in the workplace (Morgan, 2020).

"A cap of 25 percent will enable us to trial that, come back to the office in an orderly fashion and then as we get that down pat, we would like to see those numbers increase" (The Victorian Chamber of Commerce and Industry's director, 2020). "The longer the Premier keeps all the public service working at home, the harder it's going to be for those CBD small businesses" (opposition leader, Michael O'Brien, 2020).

Possible Futures

Scenarios for 'Dollars for Docklands'

For the first couple of months it is expected that Dollars for Docklands will have a relatively slow start. A contact with expertise in hotel leadership mentioned that the initiative is little known about by the public outside of the Docklands precinct. A new 'Docklands Dollars' initiative, aimed exclusively at helping bring visitors and foot traffic back to Docklands, will be launched by the DCC on 29 March 2021. With the support of the City of Melbourne and Development Victoria, the scheme modelled similar economic injection programmes seen around the world during COVID-19 by encouraging people to stay and spend in Docklands through a range of incentives. By registering through a dedicated website – docklandsdollars.com.au – participants are required to book a minimum two-night stay at a Docklands accommodation provider and spend money at local restaurants, bars and attractions to claim a rebate from the DCC. While only businesses which are members of the DCC have been eligible to participate in the programme, customers were able to claim $50 off their accommodation, as well as 25 percent off the total spent of their meal or purchase. With the entire programme being delivered by DCC and its members, local creative agency MoWorks has been charged with creating the Docklands Dollars website, while other members have been given various marketing and promotional responsibilities (Dockland News, 2021). In light of a recent economic report by the City of Melbourne revealing a near 50 percent occupancy of street-facing businesses in Docklands, DCC president Johanna Maxwell said the scheme helped businesses recover slowly (Dockland News, 2021).

However, with successful targeting and implementation, there should be more fruitful results for the service sector and entertainment sector. While the DCC is optimistic that the residents are supporting the initiative by quoting that

> Residents have not been negative about this, we have only had positive feedback from all business and community at this stage. The residents don't want to see their business community die…this would impact their lives if the services were no longer available in Docklands.
>
> (expert in the DCC)

However, the residents will be concerned if the Dollars for Docklands is too successful. It was often mentioned by the locals that they are sceptical about short-stayers.

Regretfully, the exact duration of Dollars for Docklands is not currently known, it is mentioned by DCC president Johanna Maxwell that the initiative is meant to support businesses slowly. One scenario can lead to Dollars for Docklands stopping when visitor numbers start picking up again. However, another scenario could lead to the initiative staying active due to the fact that Docklands had been struggling to generate traffic even before the pandemic. An expert in DCC also acknowledges that it will be difficult to predict how Australia will react to the future. However, the Lord Mayor of Dockland states that

> now that the workplace caps have risen to 75 percent, we look forward to welcoming more workers back to their Docklands offices. However, we also need to expand our horizons and aim to attract new jobs and industries into our city. This means supporting start-ups to become the base of new headquarters for national and international companies. The priorities for this council term are driving the economic recovery, delivering infrastructure projects on time and on budget, getting the basics right, sustainability and making our city more equitable.
>
> (Dockland News, 2021)

Scenarios that Connectivity Will Improve

Within the next six years, bike paths are going to be added to connect Docklands to CBD, West Melbourne, South Wharf and Southbank, this will allow commuters to travel more freely and easily to their offices in Docklands and in the CBD. This will mean that car dependency will decrease, with less traffic and no parking problems. Large office buildings will not have to rely on large parking areas and Docklands can focus more on parks and other greenery instead of asphalt. This will increase overall liveability for the residents as pollution will go down. Furthermore, the plans for a separated bike path to the pedestrian path will be developed, which will allow for safe walking routes as walking is an important part of Docklands due to the proximity to the CBD.

While the large office buildings will take time to fill back up with workers, connectivity will be improved to ensure safety and accessibility once the pandemic is over and new employees fill the office buildings. The improved connectivity will also benefit event organisers, from cultural events happening in Docklands to the large events in the Marvel Stadium. The Marvel Stadium being remodelled would hinder connectivity from Docklands to the CBD, however on completion Docklands would be more accessible and the overhead railways will be covered.

The Yarra River will likely see a Yarra River Authority in the next six years, because of the urgent need to have one authority control the river. This means that the development of a ferry, metro tunnel or tram bridge is likely be planned, if not made reality. Proponents have struggled for the past seven years to establish a commuter ferry.

Scenarios for the Marvel Stadium

Four different futures were identified for the Marvel Stadium: maintaining the stadium but renovating it and perhaps its surroundings; destroying it and rebuilding a new one in another part of the Docklands (in West Park); destroying it and rebuilding a new one in another part of Melbourne (east of the CBD); or changing nothing (maintaining the stadium as it is and making no renovation at all). All those possibilities are plausible according to experts and Australian newspapers. However, the scenarios for which the most data could be found were the first and second ones: maintaining the stadium but renovating it and perhaps its surroundings or destroying it and rebuilding a new one in another part of the Docklands (in West Park). If this first option is going to happen it would have several repercussions for the Docklands, the residents and different MLOs present in the area.

The renovation of the stadium as already mentioned would transform this sports arena into a community meeting place hosting a new town hall within it. This would be good for the residents as they would be able to socialise more easily, and have more activities to do in the area. This positive outcome could perhaps make the residents more 'tolerant' about the downsides of having a major stadium in the middle of their precinct, such as noise pollution generated by fans.

Furthermore, with this renovation the whole Dockland area would become more attractive, which could be beneficial for the MLO, especially the service-oriented business. The new town hall would be able to host a lot of events which could bring new visitors to the Docklands who would be willing to spend an overnight there and so benefit the hospitality industry.

A key stakeholder regarding the renovation of the stadium is the AFL as they are the owners of the stadium. Ultimately, they are the ones who will play a major role in deciding the future of the stadium. The renovation of the stadium would create hundreds of local jobs when the construction occurs. This would benefit construction business and the service businesses located in the Docklands as workers working in the constructions site will potentially have lunch every day in Dockland's restaurants or buy food in local supermarkets. If the second option is selected, all the positive effects of the first option will remain plus other advantages. If a new stadium is built it will be more modern and functional than the previous one.

The new location of the stadium is likely to be more offset, at the end of the west of the Docklands. Thus, during a day event, the residents of the Docklands could suffer less from noise pollution from the supporters. Furthermore, the removal of the stadium would allow a better connection of the Docklands with the CBD and the rest of the city. This would make the Docklands more attractive and therefore could benefit a lot of MLOs present there. Better connectivity between the Docklands and the CBD could encourage new MLOs to relocate in the Docklands as the area is cheaper than the CBD.

After the destruction of the stadium, the idea would be to build more residential buildings where the stadium was located. This can be both seen as positive and negative for the residents. On the positive side, if more people live in the area more shops, bars, restaurants will open to satisfy the new demands and therefore the residents will have more options which would be positive for the service-oriented MLOs of the Docklands as well. On the negative side, some residents could be annoyed if their district becomes more densely populated.

To sum up, with this solution the Docklands would keep the advantage of having a big infrastructure in its heart but without a lot of the downsides (noise from the supporters, an old stadium and bad connectivity with the CBD).

Scenarios of Docklands' Changed Work Behaviour

COVID-19 has shaped the working environment in the year 2020, since then most employees have been allowed to return to their offices, however many may choose to stay at home and work from home. This trend could go on and settle as the 'new' normal. This would mean that large office buildings being needed less as virtual offices are established. The south of Docklands in particular has many high-rise office buildings. Having empty office buildings is likely not because of high overhead costs, however the number of organisations within one office building could change, as could the purpose of the office buildings, some becoming new residential homes. This would mean that many apartments will likely be rentable with decreasing rent prices. Many Airbnb operators are also struggling, meaning that many landlords are looking for long-term renters instead. The large housing market will decrease rent prices. Young people and professionals are mostly flexible in their living environment and could move to different areas, depending on the rent prices, in which case if Docklands did become less expensive in housing, more young people could enter Docklands, changing its identity.

Alternatively, the working behaviour of employees could return to pre-pandemic patterns. This pre-pandemic state could also be artificially re-created, by organisations forcing their employees to come back into the office, once COVID is controlled, allowing Docklands to return to its old normal, as known before the pandemic.

Conclusion

Docklands is a young precinct, which means that many developments still need to materialise to make this precinct match the image it is given by the City of Melbourne and its residents. In the upcoming six years radical changes are expected, many caused by the pandemic COVID-19, while others are due to the quick expansion of the precinct and the many commuters and residents.

The residents will see many changes which will impact them greatly. They however also have the power to change plans and even stop developments. NGOs have caused the tram line to be put on hold, as this might disturb the view of the Yarra River. Home office working is a significant factor for huge unoccupied office buildings and inhabitants, as fewer commuters reduce traffic. It also means that some residents might not want or need to live in the city anymore, as larger housing options are more appealing and are located outside the centre of the city. Tourism is another major factor impacting everyday life including MLOs in Docklands, which is encouraged by the Dollars for Docklands initiative. Different organisations and businesses are part of this initiative and could benefit greatly from it. Major sporting and other events in the Marvel Stadium attract visitors and also have large economic benefits. Changing the exterior, repositioning it or removing it are options that are open to the AFL, each scenario would impact the MLOs in the area in different ways.

References

Australian Bureau of Statistics. (2017). 2016 Census QuickStats: Docklands. Available at www.abs.gov.au/census/find-census-data/quickstats/2016/SSC20760%23:~:text=In%20the%202016%20Census%2C%20there,up%200.2%25%20of%20the%20population.&text=The%20median%20age%20of%20people%20in%20Docklands%20was%2030%20years.

Australian Bureau of Statistics. (2021). Census QuickStats: Docklands. Available at https://abs.gov.au/census/find-census-data/quickstats/2021/SAL20766 (Accessed April 2024).

Borland, J. (2020, September 6). Tracking Victoria's job losses: There's no road to recovery without containing COVID-19. Available at https://theconversation.com/tracking-victorias-job-losses-theres-no-road-to-recovery-without-containing-covid-19-145621 (Accessed 6 September 2020).

Braun, V., & Clarke, V. (2012). *Thematic Analysis*. Singapore, American Psychological Association.

Capp, S. (2020, December 21). Melbourne is worth 7% of Australia's economy. How can it build back post-COVID? Available at www.weforum.org/agenda/2020/12/melbourne-post-covid-recovery-plan/ (Accessed 21 December 2020).

Car, S. (2022, November 30). Council's "governance review" to tackle the Docklands Act. *Docklands News*. Available at www.docklandsnews.com.au/councils-governance-review-to-tackle-the-docklands-act/ (Accessed 30 November 2022).

City of Melbourne. (2012). *Docklands Community and Place Plan.* Melbourne: City of Melbourne.

City of Melbourne. (2016). Bicycle plan. Available at www.melbourne.vic.gov.au/bicycle-plan-2016-2020%23:~:text=The%20plan%20aims%20to%20increase,7140%20contributions%20from%20the%20community.

City of Melbourne. (2021). Multicultural communities. Available at www.melbourne.vic.gov.au/about-melbourne/melbourne-profile/multicultural-communities/Pages/multicultural-communities.aspx (Accessed 8 March 2021).

City of Melbourne. (2021). Social media. Available at www.melbourne.vic.gov.au/Pages/social-media.aspx (Accessed 8 March 2021).

City of Melbourne. (2022–2023). Annual report. Available at https://mvga-prod-files.s3.ap-southeast-4.amazonaws.com/public/2024-07/annual-report-2022-23.pdf (Accessed 1 April 2024).

Curran, L. (2018). Ten cultural experiences in and around Melbourne's Docklands. *Concrete Playground*. Available at https://concreteplayground.com/melbourne/arts-entertainment/ten-cultural-experiences-in-and-around-melbournes-docklands (Accessed 15 January 2022).

Dan, A. (2021). Virtual agencies will be the new norm in a world of empty offices. *Forbes*. Available at www.forbes.com/sites/avidan/2021/01/26/virtual-agencies-will-be-the-new-norm-in-a-world-of-empty-offices/ (Accessed 26 January 2021).

Dockland News. (2021). Fighting for Docklands with dollars!. Available at www.docklandsnews.com.au/themes/user/site/dn/pdf/DN172.pdf (Accesed 2 March 2022).

Galloway, A. (2021). Analysis ranks the countries that handled COVID-19 best. Available at www.smh.com.au/politics/federal/analysis-ranks-the-countries-that-handled-covid-19-best-20210127-p56x3g.html (Accessed 28 January 2021).

Hill, M. (2020). Tram options back on the table. *Docklands News*. Available at www.docklandsnews.com.au/editions/article/tram-options-back-on-the-table_16712/ (Accessed 2 September 2020).

Lim, S. R. (2020). The top 10 things to do in Docklands, Melbourne. *Culture Trip*. Available at https://theculturetrip.com/pacific/australia/articles/top-10-things-to-see-and-do-in-docklands/ (Accessed 16 September 2020).

Mark Doman, N. S. (2020). From ghost town to green shoots. Available at www.abc.net.au/news/2020-11-06/search-melbourne-suburbs-to-see-the-rebound-after-covid-lockdown/12851854?nw=0 (Accessed 5 November 2020).

Masanauskas, J. (2015). Lord Mayor Robert Doyle: 'Etihad Stadium must be bulldozed'. Available at www.heraldsun.com.au/news/victoria/lord-mayor-robert-doyle-etihad-stadium-must-be-bulldozed/news-story/28c6526e89af148f3d8576c28636587c (Accessed 1 March 2021).

Melbourne Docklands Map (n.d.). Map data ©2025 Google. https://nl.images.search.yahoo.com/yhs/search;_ylt=AwrIekdw7EtoiWkUYBCfKgx.;_ylu=Y29sbwMEcG9zAzEEdnRpZAMEc2VjA3Nj?p=Docklands+Map&type=ie-q1000328_3012q&hsimp=yhs-001_2&hspart=ima&ei=UTF-8&fr=yhs-ima-001_2&guccounter=1&guce_referrer=aHR0cHM6Ly9ubC5zZWFyY2gueWFob28uY29tL3locy9zZWFyY2g7X3lsdD1Bd3JrUFFwbjdFdG9FMmNWUWdXZktneC47X3lsYz1YMU1ETWpFeE5EY3lOalUxT1FSZmNnTXlCR1p5QTNsb2N5MXBiV0V0TURBeFh6SUVabkl5QTNOaUxYUnZjQVJuY0hKcFpBTnVTekJPZFZZeVVGUmpkWE5VTlVRMlluUnhabGRCQkc1ZmNuTnNkQU13Qkc1ZmMzVm5ad013Qkc5eWFXZHBiZ05lYkM1elpXRnlZMmd1ZVdGb2IyOHVZMjl0QkhCdmN3TXdCSEJ4YzNSeVlYUUVjSEZ6ZEhKc2JnTXdCSEZ4YzNSeVRpWjBZ1WkhNbE1qQk5ZWEFFZFZZEY5emRHMXdBekUxTkRrdE1EWXhPVE0tP3A9RG9ja2xhbmRzK01hcCZ0eXBlPWllLXExMDAwMzI4XzMwMTJxJmhzaW1wPXlocy0wMDFfMiZoc3BhcnQ9aW1hJmVpPVVURi04JmZyPXlocy1pbWEtMDAxXzI&guce_referrer_sig=AQAAABrKsPOI44cu1A4hxJJKRpxV8jiavob7IAGC-whCX-AdRJ-TtDT1ZKRrUgCi2kyBjaKAMbZ-lyJIbJYerCifgZ6DMm06O6hYDTDE-K1r_3Y0E9rsUH63dpGzvy9gzcEyK1pvVeOGSdD71epLXl8aS8xNo61dsQkx-VMm76BgZrqx%23id=1&iurl=https%3A%2F%2Flv.maps-melbourne.com%2Fimg%2F1200%2Fkarte-docklands.jpg&action=click

Morgan, R. (2020). Nearly a third of Australian workers have been working from home. Available at www.roymorgan.com/findings/8451-roy-morgan-working-from-home-june-2020-202006290638 (Accessed 1 March 2020).

Places Victoria & City of Melbourne. (2013). Access Docklands: A strategy for the Docklands transport network (p. 33). Available at https://mvga-prod-files.s3.ap-southeast-4.amazonaws.com/public/SiteCollectionDocuments/access-docklands-strategy.pdf

Professional Scientists Australia. (2020). The initial employment impact of the COVID-19 pandemic on Australia's science workforce. Available at https://scienceandtechnologyaustralia.org.au/wp-content/uploads/2020/08/2020-Initial-Employment-Impacts-of-Covid-19.pdf (Accessed 2 March 2021).

Rachel Eddie, B. H. (2021, March 1). Once a growing office precinct, Docklands feels 'dead'. Available at www.theage.com.au/national/victoria/once-a-growing-office-precinct-docklands-feels-dead-20210224-p575c6.html (Accessed 1 March 2021).

Scanlan, S. (2017, May 5). Regular river commuter ferry service is still being ignored. *Docklands News*. Available at www.docklandsnews.com.au/editions/article/regular-river-commuter-ferry-service-is-still-being-ignored_13020/ (Accessed 5 May 2021).

Schaefer, A. (2010). Different types of business. OpenLearn. Available at www.open.edu/openlearn/money-business/business-strategy-studies/different-types-business/content-section-0 (Accessed 23 March 2021).

Schlesinger, L. (2021, February 8). Two-thirds of Melbourne offices unoccupied, Sydney less than half full. Available at www.commercialrealestate.com.au/news/two-thirds-of-melbourne-offices-unoccupied-sydney-less-than-half-full-2-1024353/ (Accessed 8 February 2021).

Schout, D. (2021, March 3). Almost half of Docklands' shopfronts vacant: report. Available at www.docklandsnews.com.au/editions/article/almost-half-of-docklands-shopfronts-vacant-report_17213/ (Accessed 3 March 2021).

Shaw, H. (2015). *Melbourne Docklands: It's a Class Remake But It Ain'tClassy*, (pp. 1–10). Melbourne, 7th State of Australian Cities Conference, 9-11 December 2015, Gold Coast, Australia.

Shaw, K. (2013). Docklands dreamings: Illusions of sustainability in the Melbourne docks redevelopment. *Urban Studies*, 50(11), 2158–2177.

Silva, K. (2020). Melbourne lord mayor faces 'extraordinary' challenge as city wakes from COVID nightmare. Available at www.abc.net.au/news/2020-10-21/melbourne-lord-mayor-faces-huge-task-in-covid-recovery/12775958 (Accessed 20 October 2020).

The University of Sydney. (2020). Australians want to work from home more post-COVID. Available at www.sydney.edu.au/news-opinion/news/2020/09/28/australians-want-to-work-from-home-more-post-covid.html (Accessed 3 March 2021).

Travel Victoria. (n.d.). Docklands: General information. Available at www.travelvictoria.com.au/docklands/ (Accessed 4 March 2021).

Tuffield, R. (2021). Fears Docklands may not recover from COVID-19 lockdown despite easing of restrictions. Available at www.news.com.au/finance/business/retail/fears-docklands-may-not-recover-from-covid19-lockdown-despite-easing-of-restrictions/news-story/809aa75a6c726aaeda50e81fc45cf79b (Accessed 1 March 2021).

Victoria Development. (2019). Docklands. Available at www.development.vic.gov.au/projects/docklands?page=overview (Accessed 30 October 2021).

World Economic Forum. (2021). Australian state says work from home is over, but employees still shun office. Available at www.weforum.org/agenda/2020/12/australian-state-work-from-hom-employees-covid-coronavirus/ (Accessed 25 March 2021).

Part II

The Probable Future

Analyse and Evaluate: Match and Probable Future

5 The Probable Future of George Town, Penang, Malaysia

Introduction

The core and buffer zones of George Town are the focus of this study. Preservation efforts are underway in these two zones, which are classified as UNESCO historic areas. UNESCO and local authorities collaborated to set strict regulations that regulate the repair, preservation and protection of sites in George Town. The core zone has strict constraints, following Special Area Plan (SAP) criteria, with new constructions limited to a maximum height of four stories. In contrast, the buffer zone is less strict in terms of adherence to standards, allowing for broader development flexibility (Mazlan, Shi & Wai, 2023).

The core zone, encompassing the historic city of George Town, is enclosed by the Straits of Malacca waterways, Lorong Love, or Love Lane, forming the northwestern border, and Gat Lebuh Melayu, Lorong Carnarvon and Lebuh Carnarvon to the southwest (Penang Global Tourism, n.d.). The buffer zone protects the core zone, bordering a stretch of sea around the harbour area, the streets of Jalan Dr Lim Chwee Leong and Gat Jalan Prangin to the southwest and Jalan Transfer to the northwest corner (Penang Global Tourism, n.d.).

The Research Scope

This research explored the dynamics of tourism development in George Town, Penang, with a focus on understanding stakeholder perspectives, and current tourism practices and market trends, particularly among medium and large organisations (MLOs) and public organisations (POs). Thereby, key trends, drivers and challenges were identified leading to insights into the probable future of tourism development in George Town, reflecting the most probable scenario based on current market trends, stakeholder actions and development initiatives planned for the next five years. Due to George Town's significance as a cultural heritage site and growing tourist destination, this research provides stakeholders with an opportunity to understand and foresee the probable outcomes of current dynamics

DOI: 10.4324/9781003570240-7

and consequences they will face, to proactively address emerging challenges and capitalise on opportunities.

The investigated area encompasses the entirety of the UNESCO World Heritage Site of George Town in Penang, Malaysia, including the core and buffer zone illustrated earlier. The research and data gathering utilised a timeframe of three and a half weeks, stretching from 4 March until 27 March 2024 concentrated in the weeks of March 11 and 18.

To fulfil the research aim and produce relevant and effective outcomes a triangulation of multiple complementary and supplementary methods, gathering both primary and secondary data, was applied. While the focus lay on qualitative data collection, quantitative secondary data also was gathered through desk research to support and/or illustrate outcomes where required. The research was conducted using mixed methods, including semi- and structured interviews with local stakeholders, walk-ins into touristic attractions, netnography and observations, supplemented by desk research. The interviews provided a significant part of the data collected, and the selected interviewees were not exclusively tourism-related in order that the research could include different perspectives that are relevant to the tourism sector in George Town. The study identifies how the current tourist markets interact within the region, and their impacts on and dynamics with local stakeholders with a focus on MLOs. The interviews totalled seven semi-structured interviews lasting on average from 45 to 60 minutes each, scheduled beforehand and held in a predetermined setting. Additionally, four unscheduled, unstructured conversations were held, where workers or employees were approached during or between their work, holding informal conversations of approximately 15–25 minutes, about certain aspects of the item list deemed appropriate at the time of approach. Participants were selected based on their expertise and roles within their respective organisations and the tourism industry in George Town.

Research Analysis

To make sense of the complex and vast amount of qualitative data gathered, thematic analysis was the approach chosen because of its flexibility and robustness in identifying patterns, themes and underlying meanings within qualitative data sets (Dawadi, 2020). To extract the information, the interviews and conversations were transcribed, partially word for word and partially analogously, while notes were taken of the observations and netnography. Through coding, the data was categorised into overarching themes which emerged during the analysis process, according to the coding scheme. This iterative process allowed for supplementary data from various sources to be integrated and thereby enrich the analysis. Overall, this process facilitated a comprehensive exploration and synthesis of insights through various sources ensuring robust findings that informed the discussion of the probable future for George Town as presented in this chapter.

The Current Landscape of George Town

History

To provide a better understanding of the current situation in George Town, it is important to record significant historical events that had an impact on the town. George Town is the capital city of Penang, Malaysia, and has an interesting history dating back to the 9th century, when it was visited by Muslim traders from India and the Middle East. The city was colonised by the Portuguese, Dutch and British, and was established as an economically successful city in 1786. George Town was named a UNESCO World Heritage Site in 2008 for its distinctive architectural and cultural townscape, influenced by over 500 years of trade and cultural contacts between East and West in the Straits of Malacca (UNESCO, 2021).

George Town's historical growth has been closely related to global events including World War II air raids and key political shifts such as Malayan independence in 1957. After 1969, the city's economy shifted from an industrial to a service-oriented one, resulting in issues of rising unemployment. Various institutions, including George Town World Heritage Incorporated and the George Town Conservation and Development Corporation, supported efforts to conserve the historic city, focusing on preserving the city's Outstanding Universal Value through urban conservation and regeneration initiatives (UNESCO World Heritage Centre, n.d.).

Tourism Developments and Trends

In terms of tourism development, George Town has developed significantly since it was nominated as a UNESCO World Heritage Site in 2008, and Table 5.1 lists tourist arrivals from 2009 until 2023. While the UNESCO designation increased tourist visits and encouraged the development of new hospitality and tourism businesses, George Town was already well-known for its diversified cultural history, historic sites and street art scene before obtaining the UNESCO title. Nowadays, tourists are still drawn to places like the Penang Museum and other exhibitions, the Cemetery, Fort Cornwallis and the Town Hall. Furthermore, the city's rich history, gastronomic options and charming small-town ambience create an attractive destination (*The Economic Times*, 2018).

Over time, George Town has strategically embraced historical tourism as an approach to promote economic growth, with developments including efforts focusing on renovating historical buildings and areas to attract more tourists. However, as mentioned by several of the respondents, there has been a conscious effort to balance the benefits of tourism with the preservation of local heritage and its cultural identity.

Despite its achievements, George Town faces obstacles, such as challenges in conserving authenticity, avoiding over-standardisation, reducing gentrification pressures and protecting cultural rights as it pursues tourism growth. According to the respondents, several initiatives have been proposed to address these concerns, including developing stakeholder consensus, stimulating historical study and establishing procedures to assure long-term tourism growth while preserving the city's unique character.

In addition, all the stakeholders consulted noted that there is an ongoing debate about the definition and attraction of 'quality' visitors, with concerns going beyond basic purchasing power (expenditure) to include responsible behaviour and cultural understanding. However, the desk research and the results of the interviews indicate that mass tourism in Penang has been a constant trend, drawing a large number of visitors to the island and will continue to do so. Initiatives and planning documents, like the Penang Masterplan (Penang State EXCO Office, 2018), are being implemented to disperse visitors outside George Town's downtown region, with the goal of reducing congestion and promoting a more equal distribution of economic advantages within the local population.

Another development that is impacting George Town according to the interviewees is the use of technology in the tourism sector, currently affecting how visitors experience the area. Technology and the internet have contributed to a shift towards customised travel experiences suited to individual interests. The use of QR codes in businesses and heritage sites enhances tourist experiences, simplifies operations and addresses labour shortages. By utilising technology for real-time feedback and changing products to match market trends, stakeholders can ensure a more dynamic and engaging tourist experience, catering to different visitor preferences (*Digital Penang*, 2020).

Features and Assets

George Town in Penang has several assets and features that emerged from desk research, and interviews, pointed out by blog posts consulted for netnography, and from observations that contribute to its active and well-connected status both within Malaysia and globally. As well as its multiculturalism, the city is known for its famous street food, with Malaysian curries, local Malay, Indian and a variety of other cuisines characterising offerings across the town. Downtown George Town's heritage appeal is visible in its historical architectural structures, handicrafts and different cultural communities such as Little India and Little China, each sharing and celebrating heritage with colourful festivals and events (Cahill, 2024). In addition to its history, George Town has a variety of leisure opportunities and alternatives, including museums, such as the Ghost Museum and Asia Camera Museum, as well as street art and murals that continue to attract visitors.

Challenges

George Town faces several challenges that impact the city's development and sustainability. One significant issue is the strain on transportation and infrastructure due to the growing number of tourists and residents, leading to congestion and inefficiencies. This challenge arises mainly because of slow development processes. This perspective was reinforced by the representative of the Tourist Guides Association who mentioned that the slow pace of implementation can be connected to bureaucratic and political processes, as 'many promises are being made,' without or with only limited follow-ups on those plans. In addition, a challenge that emerged in the study and is addressed in the aforementioned masterplan is a lack of reliable data. Obtaining comprehensive and integrated data on tourism in George Town is critical for making informed decisions, yet, current sources lack cohesiveness and may not completely represent the complex nature of visitor patterns, numbers and behaviours. Establishing a trustworthy and uniform data source is critical for correctly assessing Penang's tourist scene and supporting strategic planning activities (Penang State EXCO Office, 2018), and this lack is affecting all stakeholders involved.

The transport system infrastructure in Penang, and George Town specifically, is failing due to limited space in major town areas, rapid growth of car ownership, overcrowding of people and traffic congestion from a high volume of vehicles. The transportation system failure in Penang is attributed to limited consultation with the community, public organisations and internal-centric decision-making processes. Moreover, the rising prices of residential housing have forced many residents to relocate due to affordability issues. This is a trend that the respondents see continuing further. Once the current rent control is finished, land-owners greatly increase rents, so people cannot afford to stay in their homes.

Preserving the intangible heritage of George Town is another critical challenge highlighted by respondents. With the city's ongoing focus on tourism growth, there is a growing concern about the potential loss of its authentic charm and cultural identity. This concern is related to the trend of locals relocating from George Town, hindering their ability to engage in traditional practices within the city. Another challenge is the limited availability of land in Penang and George Town which poses a constraint on expanding tourism infrastructure and facilities, requiring stakeholders to optimise existing space and explore innovative solutions to meet the growing demands of tourism (Think City, 2021).

Attending Markets

Penang continues to attract a significant number of tourists annually. According to information from Tourism Malaysia, the state received 5.59 million tourists in 2022, an increase of 4.26 million tourists compared with 2021 during COVID-19 (The Malaysia Reserve, 2022) (see also Table 5.1). This strong recovery shows that Penang is becoming more popular as a travel destination. This growth in tourism fits well with Penang's current markets as it shows that the region remains attractive to visitors from all over the world. Based on continuous mentions by respondents, several key markets stand out as already significantly influencing George Town's tourism and are expected to dominate future markets as well.

International Visitors

Penang attracts a significant number of international tourists, especially from neighbouring countries including Singapore, Indonesia and Thailand, and also China. Tourists from more distant regions such as Europe, Australia and North America also contribute to Penang's international tourism market. The tourists entering through LTAPP, the Penang airport, arrive from more than 40 countries, with the highest number coming from Indonesia at 163,187 [CS1] arrivals, followed by Singapore (74,800), Taiwan (15,602) and China at 13,029 people (The Malaysian Reserve, 2023).

Domestic tourism also plays a crucial role in Penang's tourism sector, with many Malaysians visiting the state for holidays, family trips and weekend getaways. Many domestic travellers are drawn to Penang for its accessibility, affordability and the opportunity to explore their own country's rich history and culture. According to My (2024), this market is important for the destination as it not only helps retain money in the region but also contributes to boosting the country's economic recovery and promoting tourist spots to all Malaysians. Respondent of the Tourist Guide Association stated the following regarding the market of Penang: "The growing popularity of Penang is attributed to its tourist-friendliness, characterized by an open market that caters to all types of tourists."

This statement regarding the growing popularity of Penang as a tourist-friendly destination is supported by the diverse offerings catering to different groups of tourists. First, Penang's Cultural and Heritage Tourism offering is a significant draw, especially with its UNESCO World Heritage Site in George Town. This area boasts well-preserved colonial architecture, vibrant street art, ancient temples and informative museums, attracting visitors keen on experiencing the unique blend of Chinese, Indian and Malay cultures. Penang's cultural richness makes it a prime destination for those interested in history and heritage.

Second, Culinary Tourism plays a key role in Penang's appeal. The city is renowned for its street food scene and diverse culinary offerings, with tourists flocking to Penang's hawker centres and local food markets. The variety and quality of food make Penang an exciting destination for food tourists. Third, beaches and nature tourism are other significant attractions in Penang. The coastal areas, particularly Batu Ferringhi Beach, offer opportunities for relaxation and water activities. Finally, eco-tourism activities such as hiking in Penang National Park and exploring Penang Hill attract nature and adventure tourism.

Chinese Market

There was a notable development with the introduction in 2023 of visa-free access for tourists from certain countries, including China and India, and this is expected to significantly boost the number of tourists in Malaysia, with an estimated 20–30 percent increase from China alone. In addition, the Penang state government is in discussions with various airlines in China to increase direct flights from China to Malaysia, with Penang as a destination (Bernama, 2023). The growing

Table 5.1 Tourist arrivals (in millions)

Year	*2009*	*2010*	*2011*	*2012*	*2013*	*2014*	*2015*	*2016*	*2017*	*2018*	*2019*	*2020*	*2021*	*2022*	*2023*
Arrivals	5.96	5.99	6.02	6.09	4.70	6.85	6.35	6.38	6.38	6.66	6.39	1.33	4.26	5.59	6.97

Source: Created by the authors.

interest and active efforts to attract more tourists from China to Penang strongly indicate that China will soon become an even more important market for George Town [CS1]. According to a recent article by Bernama (2023), hotels and industry associations are intensifying their promotional efforts to attract the attention of Chinese tourists.

Cruise Tourism Market

In 2024, Penang welcomed 573,178 international passengers to its shores, an increase of 30.56 percent compared with 439,012 passengers in 2022 and 39.82 percent compared with 409,943 passengers in 2019. According to state Tourism and Creative Economy Committee chairman Wong Hon Wai, this significant recovery indicates that Penang's cruise tourism has not only returned to pre-pandemic levels but has also exceeded its previous performance. He said in 2022, the Penang Swettenham Pier Cruise Terminal received 84 international ships, showing a substantial growth of 85.71 percent in 2023, reaching 156 ships. The increase in both cruise ship arrivals and international tourists reflects the flourishing cruise tourism industry in Penang and underscores Penang's growing status as a premier cruise destination (Dermawan, 2024).

The top nationalities in terms of numbers of cruise passengers in 2023 came from Singapore, India and Indonesia, reflecting growing interest in cruise holidays across Asia. Other Southeast Asian countries like the Philippines also contributed significantly to the region's cruise popularity. Moreover, countries like China, the United States and South Korea demonstrated strong involvement in the cruise industry. This growth sets the stage for welcoming an unprecedented number of cruise tourists in the coming years.

Stakeholder Landscape

The city represents a diversity of businesses catering to different traveller preferences and budgets. There is a coexistence of micro/small businesses alongside medium/large organisations, providing a wide range of services and options. The micro/small businesses include family-owned guesthouses, boutique hotels, homestays, as well as a great number of street food vendors, small cafés and artisans offering handmade goods and souvenirs. Medium/large organisations in George Town include upscale hotels, larger restaurants, shopping malls, retail outlets and tourist attractions as museums.

These organisations' roles are to serve both locals and tourists. Upscale hotels specifically serve wealthier tourists, and some tourist attractions are important in preserving the tangible and intangible cultural assets of George Town, such as the Peranakan Museum. As the capital city of Penang, George Town hosts the headquarters of many public organisations such as the Penang State Government Offices, Penang Island City Council, Penang Tourist Information

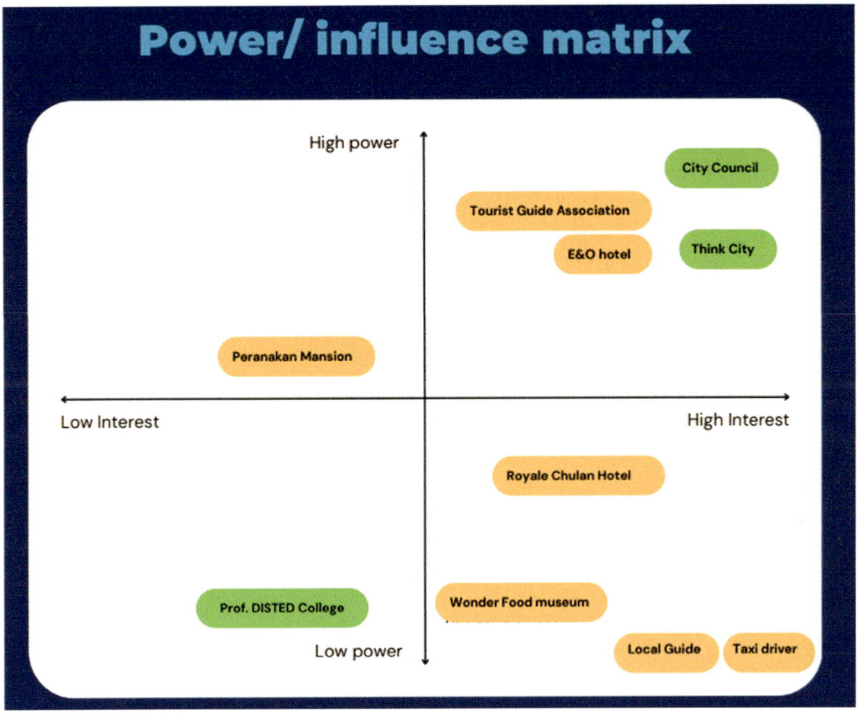

Figure 5.1 Grid power/influence matrix.
Source: Created by the authors.

Center and George Town World Heritage Incorporated, among others. These public organisations' roles are to manage and coordinate various aspects of city governance and tourism promotion, as well as to preserve and promote the city's cultural heritage assets (see Figure 5.1).

In terms of tourism development, the stakeholders approached were classified as either leaning towards an interest in increased tourist numbers in George Town or towards an interest in improving the quality of life within the city of George Town. The outcomes of interviews and observations indicated that an increase in the number of tourists is an ongoing reality and many of the stakeholders approached see this reality positively, as potentially it will result in greater revenues and thriving businesses. On the other hand, other stakeholders have a greater interest in improving the overall quality of life conditions of George Town than in increasing tourist numbers, partly because the town is currently facing difficulties in terms of infrastructure conservation. After separating the stakeholders into these two main interests, they were inserted into a grid which indicates their influence on decision-making within the context of George Town tourism development, based on their level of interest in increasing tourist numbers or improving quality of life.

Increasing Tourist Numbers

Tourist Guide Association: The Penang Tourist Guide Association (PTGA) is a labour union, created to protect the work of tourist guides of Penang. It has a high interest in increasing the number of tourists in George Town, as it would benefit directly in terms of the amount of work for tourist guides, and it also has a great influence on decision-making within the context of George Town, as PTGA has a history of representation in policy-making, pointing out the challenges and problems faced by residents and visitors.

Eastern & Oriental Hotel: The Eastern & Oriental Hotel is an iconic luxury hotel with a rich history dating back to the late 19th century, its architecture reflects the colonial heritage of Penang, standing out as one of Penang's most prestigious landmarks. It stands in the grid in a similar position as the PTGA, as it has great interest in a continuous high number of visitors to Penang reflected in guests to the hotel, as well as holding a powerful position in decision-making for tourism development because of its highly regarded position as an institution.

Peranakan Mansion: This is one of George Town's main attractions in regard to cultural diversity and heritage, and is a museum based in a mansion from the 19th century, showcasing Penang's Peranakan culture and its wealthy families' lifestyle through its artifacts, furniture, clothing and jewellery. Its influence is not as high as that of other organisations like the PTGA and E&O Hotel. Furthermore, it does not prioritise increasing tourist numbers in George Town to the same extent as those. Given its well-established status as a renowned tourist attraction showcasing cultural heritage, it doesn't need to compete with other attractions for visitors, and consequently, it lacks the same level of motivation to boost visitor numbers compared with hotels and tourist guides.

Royale Chulan Hotel: Strategically located in the heart of George Town, the Royale Chulan is a luxurious hotel that has its location within George Town's heritage area as a major selling point, making it convenient for guests to explore the city's attractions and to experience a culturally immersive experience. It has a lower score on influence on decision-making within George Town, as it is a lesser-known organisation compared with the other stakeholders, but it has a strong interest in an increased number of visitors to George Town, which would be positive for the hotel's revenue.

Wonder Food Museum: The Wonder Food Museum in George Town is a tourist attraction that celebrates Penang's rich culinary heritage and diverse food culture. It stands as a low-influence player because of its medium size, but has a relatively high interest in an increased number of tourists for the benefit of selling more tickets and increasing revenue.

Asia Camera Museum: This tourist attraction provides a museum showcasing a diverse array of exhibits, including antique cameras, rare lenses, vintage photographs and other photographic paraphernalia. It is considered to be a low-influence player given its medium size and small recognition, but it has the same high interest in an increased number of tourists as the Wonder Food Museum, as that would reflect more opportunities for increased revenue for the museum as well.

Local guide: A local guide on a tricycle rickshaw was approached and asked for their individual and local perspective as a tourism service provider of tours throughout George Town. This individual guide has a low influence, but on the other hand, has a great interest in a higher number of tourists in George Town, as it would directly impact their business.

Taxi driver: Similar to the individual local guide, an individual taxi driver was approached and their individual perspective on tourism was asked. That taxi driver also has a low influence on decision-making within the tourism context of George Town but a high level of interest in an increased number of tourists in George Town, as that would impact his business positively.

Improved Quality of Life

Penang Island City Council: The local City Council (MBPP), situated in George Town, plays a crucial role in managing and governing the municipal areas on Penang Island, including George Town. As a local authority, the MBPP has an extremely high level of power in decision-making regarding tourism development in George Town and is deeply invested in improving the overall liveability of George Town for its residents, prioritising the well-being and quality of life of the local community. Although tourism is a crucial source of income for George Town, the council aims to create a more efficient and convenient urban environment for residents, ensuring their basic needs are met before catering to the needs of tourists (Penang Island City Council, 2022).

Think City: Think City is a social-purpose organisation that focuses on urban rejuvenation and revitalisation, particularly in heritage areas and historic cities like George Town. It holds a high level of influence in decision-making because of the relevance of its projects and by being funded by the government but not as high as the City Council. Think City's 'liveability first' mindset and work culture towards preservation of the cultural heritage buildings and urban spaces positions the organisation as having a high interest in improving the liveability of George Town.

The Probable Future of George Town

Based on the forces currently defining the developments present in George Town, and the trends identified, the following section introduces and discusses the probable future of George Town over the next five years. By explaining the global and local forces driving the developments that this scenario builds on, the role and consequences for MLOs and POs within this tourism development are addressed.

The Future of George Town

In the next five years, George Town will transition further into a tourist-oriented destination, rather primarily than a place to live, driven by evolving market and stakeholder dynamics and urban development trends. In line with the emergent issue of tourism growth versus quality of life dominating the current developments and stakeholder dialogues, four specific developments represent the pillars of what is most likely to happen in George Town in the next five years as illustrated in Figure 5.2.

Tourism in George Town Will Grow

Tourism growth will be a feature of the future of George Town, underscored by the collective focus of stakeholders on attracting more visitors to the city. During stakeholder interviews, it became evident that the primary objective for MLOs and partial POs is to increase tourist numbers. With the image of being a destination that is 'friendly to all budgets,' there is no common sentiment among stakeholders about which tourist segment(s) to attract, or which will contribute in the most beneficial way to the local economy and the destination overall. As, a professor at DISTID college states:

Figure 5.2 Four development pillars.
Source: Created by the authors.

> There's a lot of push now in trying to get the numbers up and more tourists coming here.

This lack of focus on targeting a specific market segment results in an emphasis on quantity over quality, highlighted by a continuous rise in cruise tourism in particular. The perception of a positively viewed controlled mode of travelling and potential future spill-over impacts overshadows the awareness of the relatively minor local economic benefits from cruise guests' low-spending habits. However, reliance on return visits of former cruise tourists who will stay in hotels in George Town is marked by uncertainty. Such a trend is not supported by any evidence. Accommodation-providers will not experience a significant financial benefit from an increase in cruise tourism and therefore it would be wiser to focus on attracting other tourist segments, including more visitors staying overnight. These are expected to increasingly arrive from India and China, reinforced by the recent introduction of free visas.

Local tour operators and accommodations would cater to this market, along with domestic and independent travellers from Asian and European countries visiting George Town for its culinary and cultural offers. Established tourist attractions in the heritage site area would experience an influx in tourists, with organised tours mainly consisting of cruise tourists, which could create the necessity to handle crowding at those attractions during certain parts of the day. With travel globally becoming more accessible and affordable for various types of tourists, growth in visitor numbers will fuel tourism growth, raising the risk of mass tourism and associated problems in specific areas of George Town.

At the same time, the certainty of tourism growth is accompanied by uncertainty over the extent to which stakeholders will endorse and practice sustainable management of this visitor influx in George Town. While an awareness of the adverse impact that tourism could have on George Town exists, inconclusive discussions by stakeholders regarding the definition of a 'quality' tourist offering and the presence of divergent goals and strategies undermine the probability of unified action in the next five years. The absence of methods of monitoring tourist arrivals, and the questionable implementation of existing development and regulatory plans, pose significant obstacles to effective management and stakeholder alignment in the sustainable management of tourism growth, making its manifestation in the next five years unlikely.

Technology Will Transform the Way of Travelling and Tourism Operations in George Town

The advancements in technology continue to transform the way that travelling and tourism is organised and conducted in George Town. Developments as elsewhere, with developments like the QR code integration already prevalent in establishments and heritage sites across the town, not only streamline operations but also offer more modern ways of conserving and interpreting cultural and architectural heritage.

With the emerging trend among independent travellers of moving away from organised group tours and travel towards customised immersive travel experiences individually designed to fit personal preferences, the use of technology and the internet are empowering travellers to craft personalised itineraries and thereby decreasing the influence of traditional stakeholders, like tour operators, guides and information centres, in shaping tourism narratives. MLOs will depend for their survival on adapting their offerings and marketing strategies to cater to the preferences of tech-savvy travellers, while also leveraging technology to access big data and enhance visitor experiences and engagement. POs on the other hand will be able to shift towards focusing on facilitating tourism practices that align with the evolving preferences of travellers as well as enhancing online destination marketing. This emphasis on improving visitor experiences and being responsive to market trends underscores the prioritisation of tourism growth over addressing the needs of residents. Driven by the customers' and tourists' continuous adoption of new technologies, combined with the strong positive sentiment prevalent among stakeholders about the adoption of technology and ongoing implementations and integration into business operations, George Town is likely to continue growing as a destination with evolving awareness among travellers worldwide.

Change Implementation and Infrastructure Development in George Town Will Remain at a Slow Pace

It seems very likely that the slow pace of infrastructure development in George Town will remain a factor hindering the city's transition into a liveable space as well as limiting continued tourism growth. Stakeholder interviews revealed widespread acknowledgement of infrastructural deficiencies, particularly in the provision of transportation and public amenities.

> There have been plans for a streetlight for pedestrians for 10 years at one specific busy road at the Chew Jetty but no implementation can be seen.

Representative Penang Tourist Guide Association Sustainability concerns also remain relating to the strong awareness of rising temperatures as a consequence of climate change in the city of George Town.

The persistent slow pace of development projects will intensify issues of congestion, overcrowding and decreasing quality of life, both for residents and visitors. The continued focus on tourism growth, reflected in the influx of tourists arriving by cruise ships in particular, will lead to overcrowding and traffic congestion during the hours of their layover at the destination and more frequent use of taxis to reach destinations in or outside of the town. Despite efforts to address these challenges, the consensus among stakeholders is that meaningful progress is unlikely to materialise within the next five years, leaving the city ill-prepared to accommodate the increasing numbers of tourists. This may not only affect tourist satisfaction but also residents' ability to follow their daily routines. As a result, the

divergence between the needs of tourists and residents could widen. Stakeholders' efforts to promote quality of life through initiatives such as making the city more walkable and green encounter significant obstacles due to bureaucratic inefficiencies and limited resources. This presents a dilemma between stakeholders focusing on promoting tourism growth and those aiming to enhance the quality of life of residents, with the slow pace of implementation improvements undermining any ability to integrate these competing priorities effectively. Consequently, the city's transition towards a tourist destination at the expense of its quality of life is increasingly likely, as stakeholders navigate the complexities of balancing economic imperatives with environmental and social concerns.

Prices of Living in George Town Will Keep Rising

Along with gentrification, continuously rising living costs will negatively impact the affordability of living in the Old Town, continuing the trend of locals migrating to outlying areas in search of more affordable housing. The heritage site will become less of a place to live and more a place to work.

This threatens to dilute the city's cultural essence and diminish its sense of community leading to a gradual erosion of the city's social fabric and decreasing the intrinsic cultural heritage and authenticity embedded within its communities and the traditions they hold. While the heritage in George Town is kept intact through regulations for preserving inscribed buildings, as well as commercialised practices and touristic offers displaying cultural practices, questions arise about the status of the living heritage as a part of the Outstanding Universal Value of George Town. This creates uncertainty about its continued eligibility for being a UNESCO World Heritage Site, threatening the value and appropriateness of the label of UNESCO World Heritage Site.

However, due to the strong desire of all stakeholders to keep that designation because of the various benefits, including a sense of pride for residents and businesses, and the regulatory frameworks in place to ensure its continuity, the status of George Town being inscribed within UNESCO will most likely not change within the next five years. However, to accommodate the tourism growth, MLOs will have to rely on innovative solutions to expand their establishments while staying within the regulations inhibiting destruction and ensuring preservation. This could include incorporating uninhabited heritage buildings into the expansion or development of hotels, and entertainment offers, especially in the buffer zone.

Conclusion

The division and lack of cohesion among stakeholders regarding preferences and strategies are evident with no short-term resolution in sight. This gap persists due to the inherent conflict between promoting quality of life and accommodating tourism growth in George Town. As a result, the probable future of the city appears to be one where tourism expansion might come at the expense of quality of life.

Moreover, the increasingly diverse array of tourists visiting the area will have a range of impacts on stakeholders, leading to continued differing focuses and priorities. Some stakeholders may prioritise managing overcrowding or target specific tourist markets, while others may persist in tackling climate change issues and creating a liveable area. There remains, therefore, the question of what the future for George Town and its residents will be like if it becomes more of a tourist destination than a place to live.

References

Bernama. (2023). Penang looking at various promotions to attract visa-free tourists. *NST Online*. Available at www.nst.com.my/news/nation/2023/12/987002/penang-looking-various-promotions-attract-visa-free-tourists (Accessed 18 March 2024).

Cahill, A. (2024). Heritage tourism: A double-edged sword for George Town, Penang. World Footprints. Available at https://worldfootprints.com/compass/asia/heritage-tourism-a-double-edged-sword-for-george-town-penang/

Dawadi, S. (2020). Thematic analysis approach: A step-by-step guide for ELT research practitioners. *Journal of NELTA*, 25(1–2), 62–71. https://doi.org/10.3126/nelta.v25i1-2.49731

Dermawan, A. (2024). Penang's cruise tourism records robust resurgence. *NST Online*. Available at www.nst.com.my /news/nation/2024/01/1004526/penangs-cruise-tourism-records-robust-resurgence (Accessed 17 March 2024).

Digital Penang. (2020). Digital transformation masterplan 1.0. *Digital Penang*. https://digitalpenang.my /digital-transformation-master-plan/ (Accessed 16 March 2024).

Mazlan, I. D., Shi, B. S., & Wai, L. C. (2023, June). Rethinking conservation of Malaysia world heritage site livability: The case of Chew Jetty, Penang. In *IOP Conference Series: Earth and Environmental Science*, *1205*(1), 012083. IOP Publishing., https://iopscience.iop.org/article/10.1088/1755-1315/1205/1/012083/pdf

My, T. (2024) Unforgettable adventures await: Explore the top 17 places in Penang. *Traveloka*. Available at www.traveloka.com/en-my /explore/destination/penang-best-place/308660 (Accessed 11 March 2024).

Penang Global Tourism. (n.d.). Heritage zones. Available at https://my penang.gov.my / culture-heritage/heritage-zones/?lg=en (Accessed 27 March 2024).

Penang Island City Council (2022) Mission and vision. Available at www.mbpp.gov.my/en/corporate/profile/vision-mission (Accessed 1 March 2022).

Penang State EXCO Office. (2018). Penang tourism master plan 2021–2030. *GTWHI*. Available at https://gtwhi.com.my /wp-content/uploads/2022/03/Penang-Tourism-Master-Plan- 2021-2030-Report.pdf (Accessed 16 March 2024).

The Economic Times. (2018). Heritage buildings, quirky street art and delicious local food make Penang an ideal getaway. *The Economic Times*. Available at https://economictimes.indiatimes.com/industry/services/travel/heritage-buildings-quirky-street-art-and-delicious-local-food-make-penang-an-ideal-getaway/articleshow/66395400.cms (Accessed 5 March 2024).

Slabá, M. (2014) Stakeholder power interest matrix and stakeholder responsibility matrix in social corporate responsibility. The 8th International Days of Statistics and Economics, Prague, September 11-13, 2014 Available at https://msed.vse.cz/msed_2014/article/358-Slaba-Marie-paper.pdf

The Malaysian Reserve. (2023). Penang sees 138% increase in tourist arrivals at LTAPP in H1, 2023. Available at https://themalay sianreserve.com/2023/11/22/penang-sees-138-increase-in-tourist-arrivals- at-ltapp-in-h1-2023/ (Accessed 18 March 2024).

Think City. (2021). George Town World Heritage Site. *Think City.* Available at https://thi nkcityinstitute.org/GTCityInTransition/#page=1 (Accessed 14 March 2024).

UNESCO. (2021). From port city to World Heritage Site: Case study of George Town (Malaysia). Available at https://whc.unesco.org/en/canopy/georgetown-2/ (Accessed 11 March 2024).

UNESCO World Heritage Centre. (n.d.). Melaka and George Town, historic cities of the Straits of Malacca. Available at https://whc.unesco.org/en/list/1223/ (Accessed 15 March 2024).

6 The Probable Future of Gurney Drive, Penang

Introduction

Gurney Drive, divided into a 'local,' 'medical' and 'Gurney Bay' area, is the most rapidly developing part of the Malaysian island, Penang (see Figures 6.1–6.3). Gurney Drive, a road and a coastal location in Penang, has evolved from a venue for cultural gatherings and dragon boat races in the 1970s and 1980s to a multifaceted tourist and residential area (Personal Communication, Consumer Association of Penang, 2024). Initially celebrated for its vibrant food scene and as a cultural meeting point, Gurney Drive has undergone a substantial transformation aimed at enhancing its attractiveness and functionality for both locals and tourists. These changes have evoked a mixture of nostalgia and excitement among the locals and residents, as well as doubts and criticism regarding the cost and maintenance of the facilities (McIntyre, 2024). This chapter examines the probable future of Gurney Drive in the upcoming five years and analyses the consequences for medium and large organisations (MLOs). In this research, businesses are identified as a medium or large organisation according to their number of employees. Medium organisations employ 50–150 employees, while large enterprises employ over 150 employees.

The data were collected during March 2024 in all areas of Gurney Drive. Observations were made and photographs were taken to understand the area better and to identify stakeholders. Data were collected first by contacting MLOs via email and social media channels, followed by in-person contact and conversations with randomly selected residents and tourists. In total, 24 interviews were conducted with stakeholders. Besides interviews and observations, netnography was used to collect valid data and opinions on the area from digital platforms, including YouTube, TikTok, Instagram, Facebook, Google reviews and TripAdvisor. Digital platforms allow individuals to express their expectations, concerns, experiences, desires and beliefs (Bartl, Kannan & Stockinger, 2016; Bowler, 2010; Tavakoli & Mura, 2018). The collected data helped provide different opinions, thoughts and experiences of the community in Gurney Drive and its surroundings. The online community was essential to consider, as some communities that were met in person were not willing to express their views.

DOI: 10.4324/9781003570240-8

Gurney Drive is distinguished by its unique selling points of sea views for its visitors and convenience for its residents. Its current landscape is marked by its high buildings, shopping malls and ongoing constructions on the land that was reclaimed. The narrow streets and one-way lanes along Gurney Drive present a major challenge for movement in the area and its residents due to the high level of traffic congestion. The area is known for its lively evenings and nightlife along Gurney Bay, with the Hawker Centre and the park along Gurney Bay, the only large leisure park in the area.

Waves of Change: Drivers, Trends and Developments

Gurney Bay

In line with the Penang Master Plan 2021–2030 (Penang State EXCO Office for Tourism and Creative Economy, 2021), the developments at Gurney Drive aim to position it as a green, family-focused and smart urban space (Sekaran, 2023). This vision is embodied in the RM 200 Million Gurney Bay project that started in May 2022 as a Penang state government initiative, with an emphasis on green living through features like water gardens inspired by Singapore's Gardens by the Bay. The project seeks to balance urban development with environmental preservation (Sekaran, 2023). Publicly, it is being described as Malaysia's largest seafront city park, designed to draw millions of visitors globally and domestically and revitalise the tourism industry in Penang. The redevelopment of Gurney Drive, stretching from Bagan Jermal to Pangkor Road, is unfolding in phases, incorporating a blend of recreational, commercial and residential amenities, with ongoing land reclamation efforts to create the necessary space.

Phase/Priority One was developed in the south and opened for the public in early 2024, and introduced family-friendly facilities including a children's play area and a skate park, while Phase Two (north) is set to add a hawker centre, retail spaces, water gardens, water taxis and a man-made beach (Sekaran, 2023). Locals are hoping that a mega seafront project will bring back lustre to the area surrounding Gurney Drive on Penang Island. Hawker stalls that were once relocated from the seafront to the Hawker Centre will, according to the current planning, move back to the Bay in Phase Two of the project. The coast along Gurney Drive is currently undergoing a massive makeover with the beachfront transformed into a public park. It will offer a totally new experience to both locals and tourists. The green lungs within the development would offer families an ideal place to spend quality time together and enjoy the sea breeze (Sekaran, 2023). The overarching vision for Gurney Bay aims to establish it as a premier seafront destination, promising a picturesque setting enriched with Penang's famous hawker fare and a conducive environment for families to enjoy quality time.

Andaman Land

The Eastern & Oriental (E&O) development of the artificial island 'Andaman' is part of a larger development initiative aimed at enhancing the region's attractiveness

Figure 6.1 Map of Gurney Drive. Map data ©2025 Google, annotated by students on MA Tourism Destination Management, at Breda University of Applied Sciences, 2025, The Netherlands.

Figure 6.2 Development and construction. Map data ©2025 Google, annotated by students on MA Tourism Destination Management, at Breda University of Applied Sciences, 2025, The Netherlands.

Figure 6.3 Gurney Bay area. Map data ©2025 Google, annotated by students on MA Tourism Destination Management, at Breda University of Applied Sciences, 2025, The Netherlands.

and functionality as a destination, with a strong focus on sustainable living and green areas (Eastern & Oriental Berhad, n.d.). It aims to offer a mix of residential, commercial and recreational facilities and is being marketed as designed to boost tourism, provide housing solutions and improve the overall quality of life for residents and visitors (Eastern & Oriental Berhad, n.d.). The Meg Penang as part of Andaman Island is a set of serviced apartments designed for young executives and families, as well as the Arica @ Andaman, a luxury residential tower offering fully furnished units Easter Oriental (2025). Phase One of the construction will take about 15 years, and 30 years in total to develop the Andaman Phase Two (C. Teng, Personal Communication, 13 March 2024).

Technological Advancement

Penangs' fast technological advancements and large electronic manufacturing sector have contributed to its reputation as 'the Silicon Valley of the East.' Technological advancements, particularly in virtual reality (VR), are being leveraged to enhance tourist experiences in Penang, including Gurney Drive. Through the recently launched app 'Welcome to Penang,' visitors can get a preview of the state's attractions (The Sun, 2023). In the app it is possible to preview areas in Gurney Drive before visiting, such as the Paragon Mall and Gurney Plaza. In addition to that, from 2020 medical patients from Malaysia with stroke and neurological conditions have had access to advanced rehabilitation services with the opening of the Therapedic Brain and Spine Robotic Centre at Sunrise Tower on Gurney Drive (Hoay, 2020). According to the Centre's founder, James Tan, this facility is Malaysia's first to utilise robotic technology for faster patient recovery.

Medical Tourism

Penang has emerged as a leading destination for medical tourism, particularly favoured by Indonesians due to direct flights from cities like Jakarta, Medan and Surabaya (Penang State EXCO Office for Tourism and Creative Economy, 2021). The demand for medical services in Penang has grown significantly over the last decade, supported by numerous reputable hospitals. These institutions offer specialised services in various fields, including cardiology, orthopaedics, oncology, neurology, ophthalmology and dentistry. Over the past ten years, revenue from medical tourism in Penang has consistently risen. This trend underscores Penang's importance in the global medical tourism landscape, driven by high-quality healthcare services and its strategic geographic location.

According to the Master Plan, Penang not only seeks to enhance its capabilities in the healthcare sector but also to position itself as a top choice for tourists seeking medical and wellness treatments (Penang State EXCO Office for Tourism and Creative Economy, 2021). It is emerging as a leading destination for medical tourism, known for its affordability and access to internationally trained doctors, contributing half of Malaysia's medical tourism revenue (Wee, 2023). Under the banner 'Good Health Great Holidays,' Penang's medical sector is expanding facilities and broadening its international reach. Recent developments include new wings and treatment centres at key hospitals.

Singapore focuses on high-end, specialised medical treatments, with its advanced healthcare infrastructure and research capabilities (Sendingan, 2021), whereas in contrast, Penang capitalises on its affordability, appealing vacation environment and comprehensive healthcare services, making it a preferred destination for those seeking quality treatment at lower costs and the added benefit of a holiday experience. The medical expertise of Penang is of the same quality as in Singapore or Kuala Lumpur, with those destinations focusing more on cosmetic surgery for tourists in comparison with Penang.

Sustainable Development and Accessibility

Gurney Drive's sustainable development is supported by significant infrastructure projects, including the Penang Transport Master Plan (PTMP) and includes a planned bicycle lane cul-de-sac on Gurney Bay along the coastal grove. The government's subsidisation of petrol means that in Malaysia petrol is very affordable making it more convenient for people to take the car as their mode of transportation and is the primary reason behind the low demand for public transport.

The PTMP intends to enhance connectivity and reduce congestion through a comprehensive network of public and private transport options, contributing to the area's accessibility and overall attractiveness, as well as shifting residents from car dependency to public transport. The master plan aims for a planned distribution between public and private transportation, targeting a ratio of 40 percent public to 60 percent private by 2030 (Penang State EXCO Office for Tourism and Creative Economy, 2021). The waterfront is a part of the development of Gurney Bay and promises easy accessibility through 1,500 extra car park spaces, lifts, escalators, ramps for wheelchairs and strollers, cycling paths and related installations.

Environmental Considerations

Coastal erosion and waste pollution are critical environmental challenges that have influenced and motivated development strategies in the area. Efforts to mitigate erosion have led to the introduction of new green spaces and the cultivation of mangroves as a step towards sustainable development (McIntyre, 2022). The water gardens including a public promenade, seafront pier and board walks ensure more open public spaces and increased green areas. In addition to that, the developments of Gurney Bay include a new waste management and drainage system. The Gurney Bay project has sparked debate due to opposition from civil society organisations and NGOs, who argue that the extensive land reclamation involved could negatively affect the marine ecosystem, disrupt tidal flows and increase siltation (McIntyre, 2024). Despite these environmental concerns, the project received approval in January 2022 from the federal Department of Environment, allowing both the developer and the state government to proceed with the development.

Real Estate and Luxury Tourism

The transformation of Gurney Drive has attracted numerous wealthy companies and real estate investors, driving up property values (Loh, 2022). Living in Gurney Drive is expensive and attracts higher-earning locals only, with monthly rental costs ranging from 2,000 RM to 9,000 RM (€437–1966) compared with George Town where prices vary between 1,500 RM and 4,000 RM (€327–874 euro) (Interviewee Interealtor SDN BHD, 2024). The rental prices are influenced by several factors including the condition of building, its location, closeness to amenities, the size of the property and the facilities it offers. Due to its amenities, Gurney Drive is a favoured location among international expats and digital nomads.

Penang is transitioning towards attracting high-end tourists by shifting towards quality tourism, involving higher spending and longer stays (Penang State EXCO Office for Tourism and Creative Economy, 2021). The shopping malls located at Gurney Drive are partially directed towards higher-spending visitors, with a large number of luxury brands and high-end products available.

Clustering of Stakeholders

Creating clusters of stakeholders was important in order to create an overview based on the stakeholders' interest and position. The stakeholders were divided into three clusters based on their positions. These clusters are development and construction, community and living, and tourism and hospitality. Each of these clusters is unique in terms of its interest and power.

Development and Construction

This cluster includes stakeholders in all areas of Gurney Drive in which development is currently taking place and construction is present, such as the areas of Gurney Bay and Andaman Island.

Community and Living

This cluster includes stakeholders in all areas in Gurney Drive where the community has an acting position, for instance, food markets and residents visiting the food markets. In addition, this cluster also includes digital nomads or expats living in Penang or on Gurney Drive itself. The stakeholders in this cluster are important for their role in the preservation of the culture and for understanding the local lifestyle compared with the tourist markets.

Tourism and Hospitality

This cluster includes stakeholders in markets and the hospitality industry that cater to tourists, as well as facilities that cater to medical tourists such as clinics and hospitals. The hospitality sector is developing primarily in the area of Gurney

Drive, whilst the medical tourists come to Gurney Drive while benefitting from the advanced health care in Penang.

The Probable Future

Gurney Bay

In five years, the Gurney Bay developments at Gurney Drive will have undergone a significant transformation, given the aim of Penang to increase leisure over business tourism and to become a key tourism hub. The analysis of social media platforms presents an optimistic outlook for the future developments of Gurney Bay until 2029, showcasing mixed feelings about the changes. Conversations with locals at Gurney Bay reveal excitement for Penang's first park of its kind, contrasting with the mixed feelings among residents and visitors uncovered through netnography.

The data point towards a future in which the park will become a hub for sports and leisure activities, particularly enjoyed by residents. Interviews with MLOs and residents indicate growing expectations of increased local satisfaction and pride once Gurney Bay's Phase Two is completed, attracting large domestic as well as international audiences.

By the end of 2025, plans indicate that hawkers will return from the Hawker Centre to the Bay area enriching the atmosphere. The completion of all elements of Gurney Bay by 2025 is expected to draw in large number of people, since Phase One is already very popular one month after its opening to the public.

Waste Management

The introduction of a new waste management system in Gurney Bay aims to address the historical challenges of litter and uncleanliness that have impacted the area. Authorities appear optimistic about the progress in tackling waste management issues and ensuring effective maintenance. However, the presence of strong odours along Gurney Bay and criticisms highlighted in the netnography analysis regarding the hawker market's environment and behaviour raise concerns about the long-term solution of such problems. Personal observations reveal a significant amount of trash within just one month after the opening of the first phase, despite the strategic placement of numerous trash bins. This, coupled with ongoing issues of littering identified in netnography studies and the imminent return of hawkers, casts a shadow of uncertainty over the five-year outlook for waste management at Gurney Bay, with indicators suggesting the potential for recurring waste challenges. This change will conflict with the expected image of luxury and leisure.

Traffic Congestion

In the social media analyses, some scepticism about the future was voiced, particularly concerning traffic congestion. As the project completion is expected to end in

2025, when a surge in visitors is anticipated, this raises concerns about potential overcrowding.

Despite these worries, interviews consistently revealed an optimistic view that Gurney Bay will continue to attract a large number of people, suggesting that traffic congestion, an existing issue, is likely to persist. The public transportation plan, set to unfold over more than five years and to be completed in phases, indicates that traffic concerns may not only remain but could also worsen within the next five years. However, the addition of 1,500 extra car parking spaces at the waterfront will offer a partial solution to mitigate some of the congestion. The introduction of bicycle lanes along Gurney Bay is unlikely to prove effective in helping relieve traffic congestion as planned, as many residents felt the climate unsuitable for using bicycles as a means of transport as compared with casual leisure use only.

Changing Landscape

The construction and development of the artificial Andaman Island, aimed primarily at enhancing tourism and providing residential opportunities, are planned to span at least another 30 years. This timeline suggests minimal direct impact on Gurney Bay's landscape within the next five years. Meanwhile, current constructions on the bay and Andaman are transforming the sea views and landscapes, previously major unique selling points for Gurney Drive's hotels. These developments have attracted numerous international real estate investors, suggesting that within five years these investors are likely to own a significant portion of properties along Gurney Drive, particularly around Gurney Bay and on the developing Andaman Island.

Luxury Tourism

Further developments towards luxury tourism are certain to occur within the next five years. This shift is characterised by the presence, growth and developments of luxury brands and high-end products at Gurney Drive's shopping malls, as well as the numerous recently opened retail opportunities such as the Gurney Walk (2022) and hotels such as the Marriott (2024) and Ascott (2022). Although the current engagement with luxury brand stores remains limited, according to the data, there is a clear strategy to attract tourists who are willing to spend highly, and the development of new luxury hotels is aimed to further increase tourist expenditure. Residents have noted this transition, as evidenced by comments on social media and analysis, leading to concerns about the affordability and pricing of food and other essentials.

Luxury shopping malls are positively contributing to the growth and attraction of high-spending tourists. The development of medical tourism and luxury tourism simultaneously will complement and support each other and boost tourism in Penang and the economy, as wealthier medical tourists will choose to stay in the upscale hotels at Gurney Drive. This is already happening and it is therefore highly likely that the current focus of Gurney Drive on attracting luxury tourists and luxury travel will grow more heavily in the next five years.

Progress

Many residents are sceptical about the further development of Gurney Bay and its progress, however, with uncertainty about whether the planned developments will be finished on schedule. The Gurney Bay developments are planned to be finished by the end of 2025, but looking at the current progress, examining varying finalisation dates and conversations with residents, this timetable is questionable. There is no confirmation but according to Chief Minister Chow Kon Yeow, Gurney Bay will be finished as planned (Mok, 2024). In addition, Joshua Woo Sze Zeng (2024), State Assemblyman for Pulau Tikus, also recently mentioned that the Gurney Bay will be finished by the end of 2025. Any uncertainty is likely to impact the rate at which the number of tourists increase at Gurney Drive and possibly other related developments.

From Jalan Transfer to Jalan Servis

In the next five years, the local area is likely to remain relatively unchanged, with minimal development and no major projects planned. The focus in this area is predominantly on micro- and small-sized enterprises, rather than MLOs. The area is characterised by smaller hotels, local street food, small local businesses and car rental services, which do not have a significant influence on the overall development of this area. However, the development of Gurney Bay and the clinics and hospitals area, which are nearby, may well have an impact on this area. The introduction of new amenities and attractions in Gurney Bay and the expansion of medical tourism (which will need additional staff) could potentially lead to an increase in rental prices in the surrounding areas and traffic congestion.

Clinics and Hospitals

In the next five years, Penang is certain to become a leading destination for medical tourism, with a significant increase in the number of hospitals and expanded medical facilities planned. This area currently has two major private hospitals, Gleneagles Hospital Penang and Penang Adventist Hospital, which also cater to tourists coming to the area for medical reasons. In addition, the area has a variety of specialised clinics in fields such as dentistry and radiology, further enhancing its appeal to medical tourists. The focus is intended to be on improving and expanding the existing hospitals to accommodate the anticipated growing demand for medical services. The primary market for medical tourism in Penang is expected to be Indonesians, attracted by the lower costs and specialised services and treatments available. The demand for general medical treatments will grow further within the next five years reflecting technological developments, such as advanced robotics. Existing hotels in the area are strategically located around hospitals, some catering specifically to medical tourists. Simultaneously, Gurney Drive continues to attract leisure tourists drawn to its luxurious offerings. The growth in medical tourism is expected to complement the existing leisure tourism market, creating a diverse and thriving upmarket hospitality sector.

Overall, the medical area of Gurney Drive is on track to become a hub for medical tourism, with a focus on high-quality care, specialised services and strategic hospitality offerings while also catering to the needs of leisure tourists looking for a memorable and rejuvenating experience in Penang. The Penang Tourism Master Plan sees an opportunity for medical tourism to emerge and develop further in the future and aims to successfully market Penang as a medical tourism destination with a high level of hospitality. The development projects are often financed by banks, which illustrates their trust and confidence in the success of the medical tourism sector. In addition, the development of the area and Gurney Bay with its strategic location will further attract visitors combining their purpose of visiting with leisure activities (Penang State EXCO Office for Tourism and Creative Economy, 2021).

Future Markets

In the next five years, the Gurney Bay area is expected to cater increasingly to high-end visitors and residents, drawn by the recent increase in luxury hotel and condominium developments. Despite these shifts towards attracting higher-spending visitors, Gurney Bay continues to be known for its leisurely and family-friendly qualities, suggesting that local residents and families will remain key users of the park. The conclusion from observations is that locals will continue to come to Gurney Drive to enjoy the modern developments, whereas tourists appreciate the familiarity and comfort of this area.

With ongoing construction likely changing the signature sea views of Gurney Bay's hotels, the area's appeal is expected to lean even further towards attracting visitors who value shopping, convenience and a luxurious atmosphere. A common view expressed in interviews was that the artificial beach, as part of Phase Two developments of Gurney Bay, will not appeal to regular beach tourists.

The upscale development and expansion suggest Gurney Bay will increasingly attract wealthier individuals and international as well as national investors, a trend reflected in the cost of living along Gurney Drive. Other developments such as the expansion of the airport is aimed at enhancing the welcoming atmosphere, further increasing Gurney Bay's appeal to visitors with significant spending power.

Depending on their budget, Gurney Drive visitors who come to Penang for medical reasons are likely to favour the luxury accommodations along Gurney Bay, as well as the rather more affordable budget hotels behind Gurney Bay, seeking to combine the affordable yet high-quality treatments with leisure activities available in the area such as shopping, enjoying local cafés or restaurants and walking along the Gurney Bay with their families whilst getting food from the hawker stalls and watching the sunset.

Implications for the MLOs

The tourism and hospitality industry is preparing itself for an increase in high-end tourists and an overall rise in visitors to Gurney Drive. The developments along

Gurney Drive have important implications for MLOs such as hotels, hospitals, shopping malls, real estate companies and construction companies.

Hotels

Hotels are likely to continuously see benefits from opportunities for growth and expansion, driven by the reputation and further development of Gurney Bay. The expected expansion of existing hotels along Gurney Drive and the development of additional new ones will lead to heightened competition in the future, which will prompt these hotels to enhance their facilities even further and to provide notable service to ensure the loyalty to their customers. The daily average occupancy rate of the new hotels will most likely rise in the next five years, attracting more high-end, leisure and medical tourists to the Gurney Drive area. However, a future challenge includes the disruption of sea views, currently the USP of many Gurney Drive hotels, because of the developments of Gurney Bay and Andaman Island. To address this challenge, hotels will need to diversify their strategies and shift their USPs further towards convenience, shopping, luxury environments, close amenities and walkability to Gurney Bay.

Real Estate

As Gurney Drive evolves into a key centre for medical, luxury and leisure tourism, real estate owners will see an increased demand for properties in the area. This surge will provide them with the opportunity to raise property prices, especially as the area's amenities continue to improve and contribute to the growing interest in properties.

Hospitals

Gurney Drive is planned to become increasingly popular among medical tourists in the next five years, leading to an increase in the number of patients to hospitals in the area. This will necessitate not only greater hospital capacity but also increases in accommodation options for them during their treatments. Additionally, a rising number of staff is to be expected in order to manage the growing influx of patients seeking medical assistance at Gurney Drive and create additional demand for residential properties. The rise in demand will lead to further expansions of the existing clinics and hospitals along Gurney Drive.

Shopping Malls

Shopping malls anticipate an increase in customer numbers over the next five years, coinciding with the influx of tourists to Gurney Drive. The stores within these malls are ready to meet the high expectations of luxury visitors they currently already serve, and further development of more luxury brand stores is anticipated. Currently, the malls face a challenge with limited parking spaces in Gurney Drive, which could lead to insufficient available parking spots for their customers in the future, hence plans to expand the number of parking spaces in the area.

Construction Companies

In the next five years, construction companies will be busy completing projects along Gurney Drive, such as putting the finishing touches to Gurney Bay Phase Two, and once these projects are finished, they will most likely shift their focus to other developments on the island, including the construction of a cable car and further expansions. Moreover, the expected rise in visitors to Gurney Bay will necessitate increased maintenance and the development of new stores and hotels, and the development of Andaman Island is planned to continue for an estimated minimum of 30 more years.

Conclusion

Based on the current trends, netnographic analysis and interviews with local residents, MLO representatives and political figures, Gurney Drive in Penang is set to undergo significant transformation within the next five years. This shift is largely driven by the Gurney Bay development and a strategic push towards luxury and medical tourism. This period will be marked by optimism for a future in which the bay becomes a focal point for leisure and sports, despite existing concerns such as waste management and traffic congestion. The ongoing development of high-end retail and accommodation, coupled with strategic enhancements in medical facilities, positions Gurney Drive as a premier destination for both leisure and medical tourists. These developments forecast growth in the local economy and also the reshaping of the area's social and physical landscape, attracting an increased number of visitors and investors. However, in 2029, Gurney Drive will most likely have to deal with challenges such as the environmental impact of the developments, waste management efficacy and the preservation of the areas' local character.

References

Bartl, M. H., Kannan, V. K., & Stockinger, H. (2016). A review and analysis of literature on netnography research. *International Journal of Technology Marketing*, *11*(2), 165. https://doi.org/10.1504/ijtmkt.2016.075687

Bowler, G. M., Jr. (2010). Netnography: A method specifically designed to study cultures and communities online. *The Qualitative Report*, *15*(5), 1270–1275.

Eastern & Oriental Berhad. (n.d.). Andaman. Available at www.easternandoriental.com/andaman/?utm_source=google&utm_medium=sem&utm_campaign=Andaman-SEM-Website-AOM&utm_term=Andaman-SEM-AOM&gad_source=1&gclid=Cj0KCQjwwMqvBhCtARIsAIXsZpZ1Azw36NJMIehEGt-_xfZEQL8YeGRhgWV-ns_JInEarY_W7TBNxusaAkVFEALw_wcB (Accessed 13 March 2024).

Easter Oriental (2025) Between urban and island is Andaman. Available at www.easternandoriental.com/andaman/%23urbanisland (Accessed 13 June).

Hoay, K. (2020). First Therapedic Brain and Spine Robotic Centre launched in Penang. *Buletin Mutiara*. Available at www.buletinmutiara.com/first-therapedic-brain-and-spine-robotic-centre-launched-in-penang/ (Accessed 13 March 2024).

Loh, A. (2022, July 25). The changing face of gurney drive. *The Star*. Available at www.thestar.com.my/news/nation/2022/07/25/the-changing-face-of-gurney-drive (Accessed 14 March 2024).

Map data ©2025 Google annotated by students on MA Tourism Destination Management, at Breda University of Applied Sciences, 2024, The Netherlands.

McIntyre, I. (2022). Coastal erosion in Penang prompts calls for fresh hydrology study. *The Vibes*. Available at www.thevibes.com/articles/news/53466/coastal-erosion-cases-in-penang-prompt-calls-for-fresh-hydrology-study (Accessed 18 March 2024).

McIntyre, I. (2024). Penang's long awaited Gurney Bay greeted with flak from former CM. *The Vibes*. Available at www.thevibes.com/articles/news/100904/penangs-long-awaited-gurney-bay-opens-with-flak-from-former-cm (Accessed 18 March 2024).

Mok, O. (2024, February 4). Penang's Gurney Bay (Phase One) finally open to the public today (video). *Malay Mail*. Available at www.malaymail.com/news/malaysia/2024/02/04/penangs-gurney-bay-phase-one-finally-open-to-the-public-today-video/116295 (Accessed 16 March 2024).

Penang State EXCO Office for Tourism and Creative Economy. (2021). Penang Tourism Master Plan 2021–2030. Available at https://gtwhi.com.my/wp-content/uploads/2022/03/Penang-Tourism-Master-Plan-2021-2030-Report.pdf (Accessed 16 March 2024).

Sekaran, R. (2023). Bringing back glory days of gurney drive. *The Star*. Available at www.thestar.com.my/metro/metro-news/2023/06/28/bringing-back-glory-days-of-gurney-drive#:~:text=Long%2Dawaited%20project&text=Gurney%20Bay%2C%20which%20costs%20RM2 00mil,this%%20picture20taken%20in%201978 (Accessed 17 March 2024).

Sendingan, S. (2021). Singapore's medical tourism ambitions falter as Malaysia steps up. *Healthcare Asia Magazine*. Available at https://healthcareasiamagazine.com/healthcare/in-focus/singapores-medical-tourism-ambitions-falter-malaysia-steps-0

Tanjung Pinang Development. (2024). Gurney Wharf – A Penang State Project. Available at www.stp2.my/gurney-wharf-background.php

Tavakoli, R., & Mura, P. (2018). Netnography in tourism – Beyond Web 2.0. *Annals of Tourism Research*, *73*, 190–192.

The Sun. (2023). Penang launches VR experience to boost tourism: The Sun. Available at https://thesun.my/home_news/penang-launches-vr-experience-to-boost-tourism-DI10813123 (Accessed 20 March 2024).

Wee, P. (2023). Penang readies itself for booming medical tourism. Travel Weekly Asia. Available at www.travelweekly-asia.com/Travel-News/Travel-Trends/Penang-readies-itself-for-booming-medical-tourism (Accessed 16 March 2024).

7 The Probable Future of Penang Hill, Penang, Malaysia

Introduction

Penang Hill is also known as Bukit Bendera and is a popular tourist destination located in Penang, Malaysia. The area is famous for its nature, lower temperatures and historical significance than elsewhere on the island. Penang Hill possesses unique flora and fauna and is home to important heritage sites, such as the Penang Hill Railway and the Bellevue Hotel. In recent years, however, Penang Hill has been facing several challenges affecting the preservation of the area, and there has been growing concern about the impact of tourism on the environment. Considering the current challenges and opportunities in Penang Hill, especially as a tourist destination but also as a UNESCO Biosphere Reserve, the creation of a probable future is advantageous for stakeholders operating in the area. There is a need to establish a symmetry between the stakeholder's aspirations and its current reality in order to maximise the potential of the area while preserving its uniqueness. This chapter investigates the probable future of Penang Hill as well as the ramifications for medium and large organisations (MLOs).

History

Penang Hill is known as Bukit Bendera in Malay and refers to Flagstaff Hill, the most developed in a range of ten hills. It is located just six kilometres from George Town and is considered one of the must-see tourist attractions on the island.

Established in 1788, it is the first and oldest British Hill Station in Asia, used as a strategic lookout point to signal the arrival of any ship to Fort Cornwallis below. It was a popular retreat during the British colonial period to escape the hot climate. In 1923, Malaysia's first funicular railway was created to make the journey up the hill quick and easy. This spawned a boom in the development of bungalows and villas, and eventually its evolution into a primary tourist attraction on Penang Island (Penang Hill Corporation, n.d.).

The funicular has gone through many renovations over the past century, the most recent in 2011. Since then, visitor numbers have skyrocketed. In 2008 there were 0.5 million visitors compared with 1.74 million in 2018 (Bernama, 2019). The hill features a variety of attractions, restaurants cafés and accommodations. One of

DOI: 10.4324/9781003570240-9

the newest is The Habitat, a rainforest discovery centre open since 2016, which is designed as an area in which visitors can interact with nature with minimal environmental impact. A notable feature of the park is that it used existing trails and infrastructure in its creation and did not use any heavy machinery, minimising any damage to the rainforest (The Habitat Foundation, n.d.).

Recent Developments

Even since the onset of the pandemic, Penang Hill has continued to evolve. In 2021 it was designated as a UNESCO Biosphere Reserve, which requires it to contribute to the conservation, socio-economic development, education and maintenance of cultural identity and tourism (UNESCO, 2022). New attractions are being created, the most recent being the Gallery@Edgecliff, a museum showcasing the local heritage and biodiversity.

Other infrastructure developments are also making progress. The Replacement Penang Hill Special Plan 2020 is guiding these developments, a primary facet of which is the construction of a cable car service. There is also the Upper Station Upgrading Project to expand the Upper Station building with a Green Building Index Gold rating (Property News, 2021). Other current projects include accessible renovations to the Sky Walk and new viewing decks.

The Penang Tourism Master Plan 2030 includes several plans for Penang Hill. The Colonial Bungalows Project aims to turn the Hill's bungalows into boutique hotels and affordable accommodations. Eco-tourism is also an initiative in Bukit Bendera, as well as the creation of a Rainforest Research Centre (Penang State EXCO Office for Tourism and Creative Economy, 2021). Penang Hill is a unique destination with a rich cultural and environmental heritage and being a mere 30 minutes from George Town, it provides easy access to one of the oldest rainforests in the world, impressive views of the island and a variety of leisure opportunities for tourists and locals alike.

The Current State of the Area

Penang Hill is a popular tourist attraction for both domestic and foreign visitors since it is situated in one of northern Malaysia's most well-known natural and cultural tourist locations. Penang Hill welcomed 1.8 million visitors in 2019, and more than 400,000 people had already visited it in the first three months of 2023 (Cjmy, 2023). It is considered a must-see spot while visiting Penang Island, and effective management is in place for Penang Hill, with stakeholders actively involved in the improvement of the area (Salman, Jaafar, Mohamad & Malik, 2021).

Assets and Unique Selling Points

Penang Hill is known for its fusion of nature and cultural heritage, and it has one of the world's oldest rainforests. The current assets found in Penang Hill include cafés, restaurants, museums, mosques, temples, hotels, spas and bungalows.

One of the top unique selling points (USPs) of Penang Hill is the opportunity to immerse oneself in nature without venturing far from the city. Penang Hill offers breath-taking views, nature sights, wildlife, culture, heritage food and a wide range of accommodations (Penang Hill Corporation, 2021). Families with children and those with disabilities can easily navigate the park, enhancing the opportunity for all to connect with nature. As a major touristic attraction of economic and environmental importance for the island of Penang, Penang Hill has several stakeholders involved in the management and development of the region. Each of these stakeholders plays an important role in shaping the future of the area as they all have a vested interest in ensuring that Penang Hill remains a sustainable and attractive destination. Penang is home to two UNESCO sites which have contributed to the increasing demand of tourists on the island and consequently on Penang Hill.

One development that is expected to highly affect Penang Hill as a tourism destination is the implementation of a new mode of transportation up the hill. This development will serve as an alternative mode of transportation to the funicular, which has proven to be inefficient for two reasons. First, the funicular has failed in the past, leaving visitors stranded at the top of the hill; and second, in high season queues to get down by the funicular can take up to two hours.

Methods

The predictions for the probable future are based on data that has been gathered from 5th until 20th March 2024. Six different methods have been used, namely: (1) expert interviews, (2) non-expert interviews, (3) observations, (4) photovoice, (5) netnography and (6) and a survey. The information has been analysed using a thematic analysis approach. After the thematic analysis, the following themes were chosen: area infrastructure, sustainable prosperity, management challenges and visitor experience. All these themes include different coding explaining what is included in the theme.

Findings

The following section will present first the attending markets, key drivers of change, the MLO stakeholders and then move into the themes that emerged from various methods and techniques used.

Attending Markets

The Penang Hill complex attracts, yearly, around 1.6 million visitors (UNESCO, 2022). A member of the Penang Hill Corporation stated in an interview that about 60 percent of the visitors are international visitors, the largest groups of these are from China, Singapore, the United Kingdom and Thailand. The other 40 percent are domestic visitors. In 2022, a total of 4.2 million tourists visited the state of Penang through the Penang International Airport out of a total number of visitors to the

Penang state of around 5.6 million that year. In 2023 the total number of visitors to Penang grew to almost 7 million (Dermawan, 2022). Cruise tourism has grown as well after COVID-19, and in 2023 more than 570,000 tourists visited Penang on a cruise compared with 2022 when this figure was almost 440,000. Most cruise tourists have Singaporean nationality (Dermawan, 2024). Many tourist attractions in Penang are connected to nature as is Penang Hill, and according to Wong Hon Wai sustainable nature-based tourism attracts many foreign visitors to the state (Çakmak et al. 2024).

Key Trends

Key trends are identified as the following: sustainable development and nature/eco-tourism, and these drivers have a considerable influence on the destination and will be shaping its future.

Sustainable Development

Penang Hill has belonged to the 'Penang Hill Biosphere Reserve' since 2021, celebrated for its exceptional biodiversity and social landscapes, including local communities (The Habitat Penang Hill, 2021). The area is officially a learning place for sustainable development and showcases innovative ways of living in harmony with nature and how sustainable development can be supported. The areas are part of UNESCO's Man and the Biosphere programme, which supports international collaboration. Important tasks in terms of sustainable development are conservation, environmental education and showcasing best practices in sustainable natural resources management.

Nature/Eco-tourism

In recent years, Penang has experienced higher numbers of foreign visitors, explained by government funding support and the region's natural beauty. The State Tourism and Creative Economy Committee chairman explained that sustainable nature-based tourism can attract more foreign visitors since, in his opinion, nature or eco-tourism has become much more popular. Different tourism segments are emerging, which can be noticed in Penang, with attractions such as Penang Hill receiving an increase in visitors particularly interested in nature and sustainability, as recorded in the different interviews of stakeholders. The Prime Minister of Malaysia confirmed that the government will support the tourism industry in the coming years, because they see nature-based tourism specifically as a key avenue for investment and development (Çakmak et al. 2024). Since the major tourist attractions in Penang are nature-based, an increase in nature-lovers visiting the Malaysian island is expected in the coming years and should be a major driver of change.

Stakeholders

The focus of this research is on the MLOs situated in Penang Hill. Analysis of these stakeholders is done by dividing them into the public sector, the private sector and others including the visitors and locals, to note differences between the structure of the businesses and other parties involved. These include Penang Hill Corporation, under the state government, the private sector including Habitat (which is also a foundation), Bellevue Hotel, Penang Hill, The Hickory Penang Hill Hotel, Penang Hill Gallery and Teddyvill Museum.

Theme 1: Area Infrastructure

The first theme identified is 'area infrastructure.' This includes all the components that form the environment and that are related to the overall condition of the area. This theme includes the transportation services in the area, environmental attributes, the maintenance and improvement of infrastructure which serve as the codes that comprise the theme 'area infrastructure.' The high frequency of these codes in the data indicates that this theme is of importance for the probable future of Penang Hill in the next five years.

Transportation

During the coding of the data, forms of transportation and its development were often mentioned. In particular, the train to Penang Hill was described by stakeholders as well as visitors. The stakeholders commented on the train in the context of technology and crowd management. It also became clear that the crowding at Penang Hill is partly caused by the inefficiency of the funicular and what have become excessive numbers of passengers. Allen Tan, the director of the Habitat, said, when asked about the increase in visitor numbers:

> It's obviously the train. So pre this generation of the train, this train took five minutes. It actually can take three minutes. But they slow it down to go up.

In the reviews and observations, visitors mention that the waiting time in the queue is quite long. One interviewee mentioned:

> An absolute joke. Bought tickets via the website the day before, on which there was no indication of any wait times. Arrived at the station to absolute bedlam (think screaming kids, massive crowd etc) only to be told that there was a 2.5 hour wait and no further express tickets available for the day, plus a similar wait for the return journey.

Social media visitors stated "The funicular ride was pretty fast going up a steep hill; lasted about 4–5 minutes." Regarding development plans for transportation, the Penang Hill is getting a cable car. The plan has been approved and the construction of the cable car began in May 2024 and the cable car is schedule to be operational at the beginning of 2026 (Penang Hill Corporation, personal interview 18 March 2024).

Attributes of the Environment

Attributes of the environment are elements of the area such as the wildlife, scenery, views and some infrastructure. In this code, infrastructure is related to visible elements such as paths, the clarity of the signs and the difficulty or ease of getting to the area and being in the area. This code has similarities with maintenance, however, the code of maintenance is mainly from the visitor's perspective and about the state of the infrastructure. The data on environmental attributes describe what people have seen and some of their feelings included.

Replacement of Infrastructure

Replacement of infrastructure shows patterns of development plans to change the condition of the already existing infrastructure. However, professors at the University Sains Malaysia in Penang stated that it is very hard for these kind of projects to be executed since they need government funding or approval, and once the funding approval is gained, it is uncertain that the project will be implemented since the government changes every 5 years, and the new officials in power can undo any plans that the previous government agreed upon.

In conclusion, the theme 'area infrastructure' emerges as a critical aspect for Penang Hill. This theme encompasses distinct elements such as transportation, environmental attributes, maintenance and replacement of infrastructure. The data indicate that transportation, particularly the train to Penang Hill, is a key element in the destination. The development plans for the cable car on the hill highlight the efforts to improve the infrastructure, and replacement of infrastructure is also mentioned in connection with Habitat plans for renovating restaurants on Penang Hill. These findings confirm that enhancing area infrastructure is crucial for the sustainable development and visitor experience of Penang Hill.

Theme 2: Sustainable Prosperity

The second identified theme is 'sustainable prosperity.' This theme was chosen because the reason behind the influence on the probable future of Penang, in Penang Hill, is sustainable development as a driver of change, which was highlighted several times during the analysis. Codes observed during the analysis under this theme include eco-tourism, education, tackling waste and pollution and capitalism over conservation. Some of the codes, such as eco-tourism and education, were used on several occasions during the interviews with stakeholders as they were described as the pillars for most MLOs in Penang Hill to achieve their goals. According to Allen Tan, the director at The Habitat Penang Hill,

> we are working with a number of like-minded businesses and vendors who all believe in sustainability.

Also, he said,

> We focus a lot on, uh, education. So, the habitat Penang Hill is really an outdoor education centre.

Furthermore, Dr Yang Kok Lee, project chief of Penang Hill, mentioned that

> I would want to see eco-tourism destinations developed across the entire area in harmony with humans and the environment as a whole.

One issue about education was noted by one of the researchers who observed,

> The guide shared interesting insights into the flora and fauna, but I had a hard time understanding him, because of his English language skills.

and another one wrote,

> The guy on the bus was friendly but I could hardly understand his English as he had a strong accent.

Initiatives to conserve nature and create awareness have been taken in the research area. One of the codes of sustainable prosperity that emerged during the thematic analysis was the imbalance between conservation and economic development. According to Kim Theng Ng, at the Penang Hill Corporation,

> I mentioned overtourism already. We can't really reduce it because it's a tourist landmark. We try to host the tourists responsibly and focus on regenerative tourism.

Moreover, Allen Tan stated

> Of course, we have to balance, not just the needs of protecting the environment, but also the socio-economic needs of all the communities that and the people that rely on Penang Hill and for their livelihoods,

also said that

> UNESCO allows us to continue to develop our economy, but just without destroying the environment.

In conclusion, Penang Hills is working towards a more sustainably conscious society by initiating many initiatives about the importance of conservation. However, over-tourism is still a major issue and management is confronted with difficulties in terms of tourist opportunities as they strive to strike a balance between environmental protection and socio-economic needs and visitors' demands.

Theme 3: Management Challenges

Management challenges is the third theme that will be of great significance in next five years. This theme covers multiple codes. Overall, the data indicates different challenges in management for Penang Hill as well as for those stakeholders in the local area. The coding suggests that most of these challenges lie in a lack of information online or offline, where there is not (enough) information provided about clothing, behaviour and cleaning, and much of the information that is provided is not in English. Crowd management also is an important topic in terms of management challenges, in situations such as the funicular railway crowding problems.

Crowd Management

In the interviews, the issues of crowd management and partnerships were mentioned quite often. In Penang Hill, crowd management is a challenge, as stated by the manager of Habitat when asked about the issues the hill is facing:

> Definitely managing the crowds. It's a big, big problem…The cable car will be a natural limiter, just like the train is now. So even when we have a group of, say, 200 people, the train will only carry between 50 to 70 of our guests at any one time. So, we will always have a natural staggering of visitors arriving. Queuing and all that will be better. Uh, they are trying so hard to manage it.

Crowd management is a real challenge and reflects other issues such as long queues. There are plans, as the manager mentioned, about building a cable car by 2026, to increase capacity. Another factor that came up during this interview is changing the pricing, whether lower or higher. It is clear is that Penang Hill is facing a challenge with crowd management. This is supported by other interviews, such as the security guard at the Penang Hill station mentioning the current situation at the hill:

> When you have thousands of people in one place, there's always the potential for incidents to occur.

The research team noted that Penang Hill was unexpectedly busy, and long queues formed, and research has shown many visitors to the hill complaining about long queues and busy crowds:

> We queued up for close to an hour to buy fast lane tickets only to be told they were all sold out. The place was horribly crowded. We ended up buying tickets online to visit a couple of days later.

Theme 4: Visitor Experience

This theme is focused on the experiences of the visitors to the research area. These include local residents and domestic and international travellers. Most of the information on this theme was generated through interviews, observations and netnography. The experiences identify what is going well in the area and what is missing or could be improved upon. Based on the findings, it is possible to speculate if these topics of interest might change in the future. This theme is composed of several codes: one was pricing – for multiple things, for example, entrances, transport and activities. There are also price differences by visitor segment, depending on who is visiting the area. One-time visitors were another code, as many visitors (cruise passengers in particular) only stay in Penang for a short period, and many people visit the area only once or twice. Many stakeholders would prefer visitors who stay for a longer duration.

Pricing

In the interviews it was acknowledged that the prices for tourists are higher than the ones for locals, as is the case at Penang Hill.

> So, the foreigners subsidise the locals because the locals are so cheap. Right. It's artificially suppressed the pricing. But the foreigner rate is so much more.
>
> –Allen Tan

One stakeholder explained that there is some reasoning behind the difference in price. They want a higher value tourist (read higher-spending tourist) who thus makes more money available to spend on conservation and other projects.

> It will bring, I believe, a higher value tourist, willing to spend and able to spend more money. Up to the hill because they will be priced commercially as a result.
>
> –Allen Tan

Other experts said that when people value what they offer they are willing to pay for it, and in the case of Penang Hill he is right. In this area not many people complain about the prices, only about the train and the waiting lines up the hill.

> People do value what we have to offer and they are willing to pay.
>
> –Allen Tan

But one foreign visitor was concerned about price and the different rates for locals and visitors:

> The entrance fee is expensive here.
>
> –Interviewee

> It's now 10 ringgit for locals and 50 ringgit for foreigners, 5 TIMES MORE! This is an absolute joke.

First-Time Visitor

The stakeholders mentioned that they would like people to stay longer in the area, as then they would spend more money on their overnight stays, food and activities.

> We'd love to attract people who will come to Penang to stay and to spend multiple days experiencing Penang.
>
> –Allen Tan

Other experts stated very clearly that visitors were only here for a short while, and weren't often planning on returning to the area. As some international visitors stated

> No I am not returning, because I am only staying in George Town for a short while.

Others visitors stated

> No we are not coming back because we only have a short stay in Penang.

The Probable Future of Penang Hill

The future for the MLOs in Penang Hill is bright. Development plans for the cable car have been approved by the government, and its construction has started in the second half of 2024 and it will enable an increase in visitors to Penang Hill, as well as an increase in diverse types of visitors since the cable car will be promoted to multiple audiences. This connects with the theme of 'area infrastructure.' One of the major businesses on Penang Hill, Habitat, will be an even more important stakeholder, showcasing nature conservation, eco-tourism and educational components about protecting the environment and wildlife and the history of the hill, resonating with the theme of 'sustainable prosperity' and the key element of change 'nature/eco-tourism.' Education programmes of Habitat will increase as Habitat is actively developing conservation programmes, working together with universities. Moreover, they have one of the biggest attractions on the hill, akin the eco-park which includes a Tree Top Walk and Langur Canopy Walk, displaying the diverse nature on the hill as well as a panoramic view of Penang. The energy Habitat puts into protecting its area on Penang Hill will be maintained as its management acknowledges the importance of preserving its environment for the well-being of the flora and fauna and also for people to be able to enjoy nature in a fun and relaxed manner, and thus puts funding into creating a safe and relaxing space for both humans and wildlife, which links to 'sustainable prosperity.' As for other businesses on the hill, the hawker centre will be restored and a new addition to a restaurant will be opened, also connecting to the theme 'area infrastructure.' This will boost Penang Hill as a tourist destination since more visitors going up the hill will get the opportunity to enjoy their meals as there will be more spaces to eat. Regarding pollution, an important code in the theme 'sustainable prosperity,' current effective management will continue in the next five years and the park currently does not have any visible littering, as confirmed by the absence of complaints about littering on online platforms and observations.

Regarding the paths on the hill and to the hill, the Penang Hill Corporation Association puts in a significant amount of effort to keep the paths safe and walkable, and the stable paths ensure that multiple audiences can safely make use of them. Although there are many developments on Penang Hill having positive effects, there are also some effects that need to be considered for the future. Due to the increase in visitors, which will have economic benefits, one effect will be an even bigger challenge of crowd management. Penang Hill's current situation in terms of crowd management is already quite a problem since reviews have shown many complaints about crowding. During the next five years, this will only increase as more people commute to the top of the hill. With the addition of the cable car, Penang Hill will have two main entry points for visitors, the other main entrance being the station of the funicular.

Regarding the pricing, it is certain that the current fees for the funicular will not change in the next five years and thus international tourists will keep on being charged a fee that is five times higher than the fee for Malaysian residents, which can influence the visitor experience. However, the amount of the fee that will be charged for the cable car is still uncertain as no information regarding this topic has been given yet. Another challenge is the balance between conservation and economic developments in the area, since Habitat has very different priorities than other businesses on Penang Hill. In the next five years, businesses will need to work together to make sure there is an appropriate balance between their priorities rather than an imbalance. Since Penang Hill is part of the UNESCO Biosphere Reserve, conservation goals should be met, which makes it even more important to keep a balance. A critical issue is 'sustainable development,' highlighting the need for a more sustainable approach towards development. Nature is not considered a stakeholder in every aspect of Penang Hill. How this balance will turn out relies heavily on the partnerships between the businesses involved in the area. This also challenges the wishes of Habitat, wanting to attract more nature/eco-tourism and longer-staying visitors to Penang Hill, since other businesses will attract different kinds of markets because of different priorities and supply. Therefore, five years from now this balance between conservation and economic development will be uncertain. To conclude, the future of MLOs in the UNESCO Biosphere Reserve is complex. Penang Hill's development, like a new cable car system, offers economic growth and tourism but also requires maintaining ecological balance.

Conclusion

The goal of this research was to analyse the developments in the destination from the perspective of MLOs in Penang Hill. The insights from visitors and locals were also gathered because these can pressure the MLOs to change things in the future. The research questions were (1) to what extent are the current features, assets and the potential of the reserve a match for the current markets?; (2) what is the probable future of Penang Hill in the next five years?; (3) what are the consequences for the MLOs that are connected to the area?

At the moment on Penang Hill, the features, assets and potential connect well with the wishes of the current market, except that the train cannot handle the number of visitors who want to visit the attraction. In the future, a cable car will be constructed that may solve this issue to some extent. The issue then is that the space on top is limited. Managing the visitors at the top will become more of a challenge. The number of tourists in Penang is growing every year, and because of the accessibility of the hill and its closeness to George Town, it is going to continue to attract more visitors, meaning more people are going to visit the Biosphere Reserve. Increased funding from more visitors will allow the reserve to do more projects and educate more people but raises the problem of competing interests and developments of a different type with the potential for imbalance in priorities.

References

Bernama. (2019, January 19). CM expects a record number of visitors to Penang Hill this year. *New Straits Times*. Available at www.nst.com.my/news/nation/2019/01/452367/cm-expects-record-number-visitors-penang-hill-year (Accessed 15 March 2023).

Çakmak, E., Isaac, R.K. & Nair, V. (2024) four tourist zones in Penang: Suggestions to secure their future. Penang Institute. Available at https://penanginstitute.org/publications/issues/four-tourist-zones-in-penang-suggestions-to-secure-their-future/ (Accessed 1 June, 2024)

Cjmy. (2023, April 10). Penang Hill Festival returns for its fourth edition. *Citizens Journal – Community, Food and Lifestyle*. Available at https://cj.my/129076/penang-hill-festival-returns-for-its-fourth-edition/ (Accessed 12 March 2023).

Dermawan, A. (2022). Penang Hill cable car project is on, says CM. *New Straits Times*. Available at www.nst.com.my/news/nation/2022/12/860746/penang-hill-cable-car-project-says-cm (Accessed 16 March 2023).

Dermawan, A. (2024). Penang's cruise tourism records robust resurgence. *New Straits Times*. Available at www.nst.com.my/news/nation/2024/01/1004526/penangs-cruise-tourism-records-robust-resurgence (Accessed 16 March 2023).

Penang Hill Corporation. (2021). A collection of Penang Hill upgrades. Available at https://corporate.penanghill.gov.my/index.php/en/about-us

Penang Hill Corporation. (n.d.). About Us. Available at https://corporate.penanghill.gov.my/index.php/en/about-us (Accessed 5 March 2023).

Penang State EXCO Office for Tourism and Creative Economy. (2021). Penang Tourism Master Plan 2021–2030. In *George Town World Heritage Incorporated* (No. 978-967-19661-0–5). Available at https://gtwhi.com.my/wp-content/uploads/2022/03/Penang-Tourism-Master-Plan-2021-2030-Report.pdf (Accessed 12 March 2025).

Property News. (2021, March 20). Penang Hill upper station upgrading project. *Penang Property Talk*. Available at www.penangpropertytalk.com/2021/03/penang-hill-upper-station-upgrading-project/ (Accessed 16 March 2023).

Salman, A., Jaafar, M., Mohamad, D., & Malik, S. (2021). Ecotourism development in Penang Hill: a multi-stakeholder perspective towards achieving environmental sustainability. *Environmental Science and Pollution Research*, *28*(31), 42945–42958.

The Habitat Foundation. (n.d.). The Habitat Penang Hill. Available at https://habitatfoundation.org.my/the-habitat-penang-hill/ (Accessed 8 March 2023).
The Habitat Penang Hill. (2021). The Habitat Penang Hill. In *The UNESCO Penang Hill Biosphere Reserve*. Available at https://thehabitat.my/the-unesco-penang-hill-biosphere-reserve/ (Accessed 19 March 2023).
UNESCO. (2022, June). Penang Hill Biosphere Reserve. Available at https://en.unesco.org/biosphere/aspac/penang-hill (Accessed 16 March 2023).

Part III

The Preferable Future

Consult and Advise: Match and Preferable Future

8 The Preferable Future of Old Town, Chiang Mai, Thailand

Introduction

Chiang Mai, often referred to as the 'Rose of the North,' is one of Thailand's most historically and culturally significant cities. Founded in 1296 as the capital of the Lanna Kingdom, Chiang Mai has evolved into a major tourism hub, attracting visitors with its well-preserved Old Town, rich Buddhist heritage and vibrant local culture (Mostafanezhad & Promburom, 2018). Unlike Bangkok's urban sprawl or Thailand's southern beach destinations, Chiang Mai offers a distinct blend of history, tradition and modernity, making it a unique destination for both heritage tourism and contemporary travel experiences. While tourism has long been a driving force behind Chiang Mai's economic growth, its unregulated expansion has led to concerns regarding over-commercialisation, cultural dilution and environmental degradation (Choe & O'Regan, 2020).

At the heart of the city lies the Old Town, a 1.5 km^2 district enclosed by remnants of ancient city walls and a moat. This area serves as a living museum, showcasing centuries-old temples, artisan markets and traditional wooden houses alongside boutique hotels, cafés and digital nomad hotspots. Due to its cultural and architectural significance, Chiang Mai's Old Town was nominated as a UNESCO World Heritage Site in 2015 (UNESCO World Heritage Centre, 2015), but as of today that designation still has not been approved. Rapid tourism development, globalisation and modernisation pose considerable challenges to the preservation of its authenticity and sustainability.

This chapter explores the preferable future of Old Town, Chiang Mai, addressing the pressing challenges, contextual issues, stakeholders' aspirations and strategic interventions required to sustain the city's unique character. By engaging with key tourism stakeholders, historical analysis and future projections, this study provides a comprehensive vision for Old Town's development, examining whether economic growth aligns with cultural preservation, environmental responsibility and community well-being.

DOI: 10.4324/9781003570240-11

The primary objective of this chapter is to forecast a preferable future and address implementable recommendations for Old Town, Chiang Mai. The chapter explores answers to the following key questions:

1. What are the major challenges and risks facing Old Town's tourism sector?
2. How can tourism stakeholders – including local businesses, policymakers, residents and international investors – influence the city's future development?
3. What strategic interventions can help Chiang Mai balance economic growth with sustainability and heritage conservation?

Through an interdisciplinary approach that incorporates historical analysis, stakeholder engagement and future scenario planning, this chapter provides actionable recommendations for local authorities, businesses and community groups seeking to preserve Old Town's authentic character while ensuring its long-term viability as a tourist destination. Forecasting the preferable future of destinations such as Chiang Mai's Old Town is essential for following three main reasons.

Balancing Modernisation with Heritage Conservation

As Chiang Mai continues to modernise, traditional Lanna architecture, customs and lifestyles are at risk of being overshadowed by global consumer trends and mass tourism infrastructure. Many historic buildings have been converted into hotels, cafés and commercial spaces, catering primarily to tourists rather than preserving their original functions (UNESCO, n.d.). Without strategic planning, Old Town risks losing its cultural authenticity and becoming another over-commercialised tourist hub.

Managing Tourism's Impact on the Local Community

Tourism has brought economic prosperity to Chiang Mai, but it has also led to gentrification (see Chapter 9, Nimman area), rising property prices and the displacement of local residents. Many traditional markets and local businesses struggle to compete with foreign-owned enterprises that primarily serve international visitors. Moreover, tourism dependency creates economic vulnerabilities, as seen during the COVID-19 pandemic, when travel restrictions led to severe economic downturns in the city (Ploysri, 2023). A preferable future forecasting helps destinations strategize how to focus on economic diversification to ensure resilience against such shocks.

Addressing Environmental Challenges

Chiang Mai faces severe environmental issues, particularly during the burning season (December–April), when planned agricultural fires contribute to hazardous air pollution levels (Beaulieu, Leblond, Déry & Cao, 2023). In recent years, Chiang Mai has consistently ranked among the world's most polluted cities during this period, with impacts on the health of both residents and tourism arrivals. Sustainable urban planning, green infrastructure and eco-tourism initiatives are essential to mitigating these environmental threats. By anticipating these challenges and opportunities, this chapter contributes to evidence-based decision-making that ensures a sustainable and culturally rich future for Chiang Mai's Old Town.

Chiang Mai: The Legacy of the Lanna Kingdom

Chiang Mai, meaning 'new city' (Çakmak, Tucker & Hollinshead, 2021), was founded in 1296 AD by King Mengrai as the capital of the Lanna Kingdom, also known as the 'Kingdom of a Million Rice Fields' . The kingdom encompassed much of present-day northern Thailand, along with parts of Myanmar, China and Laos. Owing to its strategic location, Chiang Mai functioned as a vital trading hub, especially between Myanmar and China. The city's history is marked by political and military struggles. In 1558, Chiang Mai fell to Burmese forces, leading to over two centuries of alternating control between Burma and the Kingdom of Siam. By 1892, the Lanna Kingdom was fully integrated into Siam, and in 1932, Chiang Mai became an official province of modern Thailand (Visit Chiang Mai, 2020).

Despite these transitions, Chiang Mai has retained its distinct cultural identity, particularly within the Old Town. Enclosed by a moat and remnants of its 18th-century defensive walls, the Old Town preserves its historical essence through well-preserved temples, traditional markets and centuries-old architecture. Today, Chiang Mai is recognised as Thailand's second-largest city with a population of roughly 1.6 million people, which makes it appealing to labour migrants relocating from Hilltribe communities and Myanmar to the city (Çakmak, Lie & McCabe, 2018).

Economic Context of Chiang Mai

Thailand, formerly known as Siam, has a long and intricate history as a kingdom influenced by shifting borders and diverse rulers. The nation's economic trajectory experienced a remarkable transformation in the late 20th century. From 1960 to 1996, Thailand's economy grew at an annual rate of approximately 7.5 percent, advancing from a low-income to an upper-middle-income country and significantly reducing poverty levels (World Bank, 2024). However, economic growth has slowed in recent decades, with challenges arising due to the COVID-19 pandemic. By 2024, Thailand's GDP was approximately 545 billion USD, reflecting gradual recovery efforts (Trading Economics, 2025).

Demographically, Thailand faces a shifting population structure. While a significant portion (15 percent) of its 72 million inhabitants is under the age of 15 years, the country also has a growing elderly population. A declining fertility rate, combined with an ageing demographic, poses potential labour shortages in the future (Statista, 2024). These demographic and economic shifts influence regional development, including Chiang Mai.

Political and Social Landscape in Chiang Mai

In addition to economic and cultural influences, Chiang Mai's political landscape has played a role in shaping its development. The 2023 elections in the country, traditionally a stronghold of the Pheu Thai Party, yielded unexpected results. The Move Forward Party (MFP) gained dominance, altering local governance priorities. MFP's unexpected win in Chiang Mai redirected local governance towards progressive aims, emphasising increased public engagement, decentralisation and environmental sustainability and moving away from conventional top-down development in favour of initiatives driven by grassroots efforts. For instance, MFP-aligned officials have advocated for the election of the governor of Chiang Mai to enhance local autonomy. Furthermore, they have called for decisive action to address the city's annual haze crisis, which includes proposing the drafting of a Clean Air Act and criticising the insufficient funding allocated for anti-smog initiatives.

This shift signalled a change in voter priorities, particularly in urban planning and cultural resource management, both of which are key to the city's tourism economy. In the 2025 Provincial Administrative Organization (PAO) elections, Pheu Thai, supported by former Prime Minister Thaksin Shinawatra, aimed to regain control. The PAO plays a crucial role in determining local development strategies, including policies related to tourism infrastructure, conservation of historical sites and regulations affecting businesses in the Old Town. Chiang Mai's political shifts are influencing tourism development, particularly in heritage preservation and infrastructure planning. Changes in political leadership can influence investment priorities, impacting sustainable tourism initiatives, environmental policies and heritage site management.

These developments highlight the dynamic and competitive nature of Chiang Mai's political environment. The interplay between national and local political forces continues to shape the province's governance and policy directions. As Chiang Mai progresses, the Old Town remains central to discussions on cultural preservation, tourism-driven economic growth and urban planning. Balancing modernisation with heritage conservation is essential to ensure a sustainable and preferable future for this historically significant area.

As Chiang Mai continues to attract both international and domestic visitors, political stability and governance will influence its capacity to balance modernisation with cultural preservation. Policies surrounding urban planning, tourism regulation and economic support for local businesses will determine the sustainability of the Old Town's tourism economy. Effective governance and strategic collaboration between national and local authorities will be essential in maintaining Chiang Mai's historical identity while fostering economic opportunities for local communities.

The Cultural and Economic Assets of the Chiang Mai's Old Town

Chiang Mai's Old Town is a historical and economic hub, attracting both local entrepreneurs and international visitors. The area is home to numerous 13th and

14th-century temples, including Wat Phra Singh and Wat Chedi Luang, which serve as major tourist attractions (UNESCO World Heritage Centre, 2015). In addition to these religious sites, the Old Town hosts a variety of small businesses, including boutique hotels, cafés, artisanal craft shops and museums, all contributing to its vibrant tourism economy.

Tourism and the service sector dominate the economy of the Old Town, with hospitality, dining, shopping and wellness services – such as traditional Thai massage facilities – constituting key industries. The area remains visually and structurally distinct from the rest of Chiang Mai due to its partially preserved city walls and ancient gates, which define its historical character boundaries. The Old Town not only serves as a cultural landmark but also as a centre for contemporary entrepreneurial activity, merging heritage with modern commerce.

Old Town's Key Districts and Their Tourism Dynamics

The Old Town is not a single, uniform area, but rather a collection of distinct districts, each with its own character and function (see Figure 8.1). Local residents and city maps reflect this segmentation, which is also evident through on-site observations. The area surrounding the Three Kings Monument, flanked by several museums, constitutes a key tourist zone due to its historical significance. Similarly, Tha Phae Gate serves as another focal point, featuring a high concentration of

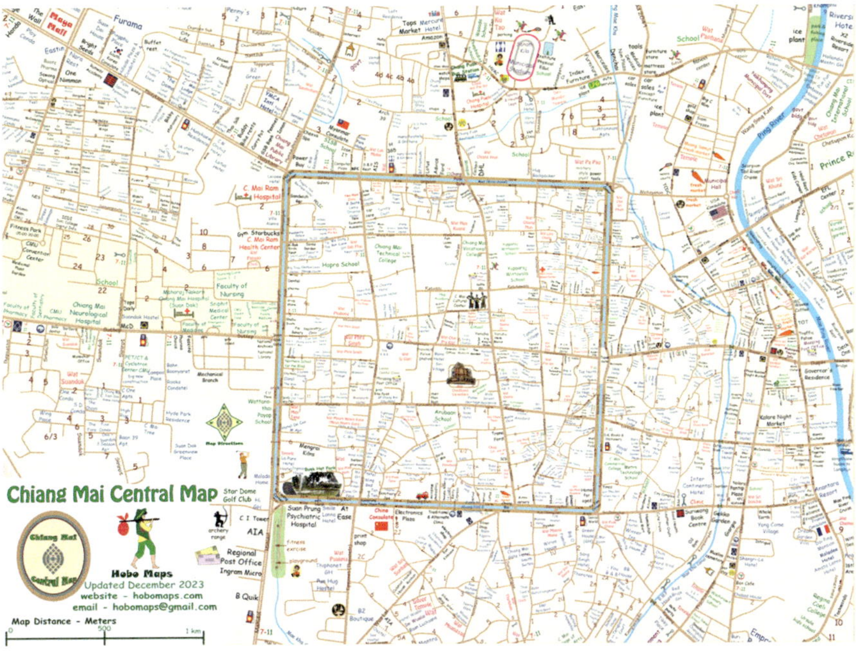

Figure 8.1 Map of Old Town, Chiang Mai (courtesy of Hobomaps). Chiang Mai Central Map.

businesses catering primarily to Western and Asian tourists. This region is also home to the Sunday Night Market (Walking Street Market), a major weekly event regarded as a must-visit attraction for visitors.

Religious sites remain the defining feature of the Old Town, attracting both international tourists and a significant number of domestic travellers and pilgrims, thus reinforcing Chiang Mai's cultural and spiritual allure. However, although the museums near the Three Kings Monument are marketed as tourist attractions, they typically draw few visitors. The area is fairly quiet, as the majority of patrons consist of school groups from Chiang Mai and other parts of Thailand.

Northeast Quadrant: The Nightlife and Entertainment Hub

This area serves as the epicentre of Chiang Mai's nightlife, boasting a dense concentration of bars, pubs and clubs that cater to young travellers, particularly from Europe, Australia and North America. The vibrant atmosphere here extends well into the night, with many venues remaining open past midnight. Although this district thrives on social and entertainment tourism, it also attracts food enthusiasts, as numerous restaurants and street food vendors operate alongside the nightlife establishments.

West Old Town: Institutional and Cultural Centre

The western part of the Old Town is home to key public institutions, including universities, schools, museums and government offices, which create a more formal and structured environment compared with the north-eastern quadrant. However, it also houses some of the city's most esteemed cultural landmarks, such as Wat Phra Singh, a temple that exemplifies traditional Lanna architecture and contains important Buddha images. This area is popular among both tourists and locals seeking to explore Chiang Mai's rich cultural heritage. Despite its institutional presence, it maintains a vibrant atmosphere with numerous small guesthouses, boutique hotels and artisan shops.

Central Old Town and Tha Phae Gate: The Cultural and Commercial Core

The central part of the Old Town is the most vibrant, serving as a major hub for commerce, tourism and cultural activity. Tha Phae Gate, one of the city's most iconic landmarks, serves as a gateway to Chiang Mai's Old Town and hosts numerous festivals, performances and public events. This area is also home to the renowned Sunday Night Market, where hundreds of vendors line the streets selling handicrafts, textiles, street food and souvenirs. The presence of numerous temples – many dating back to the 13th and 14th centuries – enhances the historical significance of the area. The central quarter plays a central role in the city's Songkran (Thai New Year) celebrations. Visitors to this district enjoy a blend of heritage, shopping and lively street life, making it one of the busiest and most visited sections of the Old Town. It is the most photogenic area of Chiang Mai, where many domestic and Asian visitors flock during both day-time and night-time to capture the best photos of their visit to Chiang Mai.

Southern Old Town: A Blend of Heritage and Hospitality

The southern part of the Old Town is emerging as a crucial area for both leisure and MICE (meetings, incentives, conferences and events) tourism, offering a rich variety of accommodations, cultural attractions and entertainment venues. The western side is famous for its boutique hotels, guesthouses and artisan workshops, which traditionally serve budget-conscious travellers. However, as Chiang Mai expands its MICE sector, these unique accommodations are increasingly drawing business travellers and incentive groups seeking immersive cultural experiences. Boutique hotels in this area offer a more intimate alternative to large convention hotels, making them perfect for small corporate retreats and executive meetings. Furthermore, traditional craft workshops and local galleries provide hands-on activities, such as artisan-led weaving or pottery sessions, which are ideally suited to incentive tourism programmes.

The eastern side of the district presents a contrasting atmosphere, boasting a vibrant nightlife that includes bars, live music venues and casual dining establishments. This area, with its vibrant atmosphere, also represents the Red Light District of Chiang Mai.

Northern Old Town: Temples and Traditional Markets

The northern section of Chiang Mai's Old Town is celebrated for its rich cultural heritage, historic temples and traditional markets. In contrast to the more commercialised central district, this area retains a strong sense of authenticity, defined by narrow alleys, wooden houses and a slower-paced atmosphere. It continues to serve as a focal point for both spiritual tourism and local economic activity, offering insight into Chiang Mai's Lanna cultural heritage.

The area is home to some of Chiang Mai's most significant religious sites. These temples are particularly notable during festivals such as Loy Krathong and Yi Peng, when their courtyards are illuminated by candlelight and lanterns, creating a visually captivating spiritual atmosphere.

Despite the growing number of visitors, the northern Old Town remains less commercialised than other areas of Chiang Mai, preserving its traditional character. Numerous accommodations in this region, including guesthouses and boutique hotels, emphasise sustainable practices and traditional Lanna aesthetics. As tourism continues to impact the district, balancing economic benefits with cultural preservation presents a significant challenge. The area's architectural integrity and religious significance require careful management to ensure that modernisation does not erode its historical value. The northern Old Town remains an essential link to Chiang Mai's past, functioning as both a religious and a cultural centre. Its blend of heritage sites, local markets and preserved urban fabric makes it a noteworthy destination for visitors in search of an authentic experience of Lanna's history and traditions.

Methodology

A qualitative research approach was adopted to explore the practices of tourism stakeholders in the Old Town. The research was grounded in ethnographic fieldwork, collecting data on informants' perceptions, thought processes and actions concerning their daily lives, professional activities, social interactions, aspirations and visions for the future. Fieldwork was conducted over six periods between 2014 and 2025, employing a combination of participant observation, interviews and document analysis. Data collection involved observing tourism stakeholders in their work environments, engaging in tourist activities to experience interactions with tourism professionals first hand, participating in social activities with informal tourism entrepreneurs, and conducting in-depth interviews and informal conversations. Additionally, online visitor reviews from social media and blogs were analysed, alongside tourism master plans, public policy documents and promotional materials from tourism entrepreneurs. Photographic documentation and personal narratives further enriched the dataset.

Narratives were collected through semi-structured, in-depth interviews focusing on informants' work histories and institutional changes. An interview guide was used, covering broad themes such as professional background, occupational shifts and industry transformations. A total of 62 informants were recruited through purposive and snowball sampling techniques. The sample included tourism entrepreneurs (e.g. homestay-owners, transport providers, street vendors, tour operators, museum and temple clerks, café- and catering-owners), as well as tourism scholars, historians, tour guides and NGO representatives. Informants were diverse in terms of age (18–64 years), gender (36 women, 26 men), ethnicity (Thai, Burmese, and members of Northern Thailand's hill tribes), religion (Buddhist, Christian, Hindu, Muslim), socio-economic background and geographic origin. Interviews, conducted in English, lasted between 35 minutes and 4 hours and were primarily held at informants' workplaces. Given the longitudinal nature of the study, multiple interactions with informants were common, and some interviews took place at different locations.

A thematic content analysis was employed to systematically examine the data. First, field notes and visual data were reviewed and analysed. Next, interviews were transcribed into MS Word documents and subjected to an initial coding process using NVivo 10. The data were then deconstructed into categories based on recurring patterns and key themes. Through axial coding, keywords, similarities and differences were clustered into thematic groups, offering insights into informants' aspirations and the broader socio-cultural and economic contexts of tourism in the Old Town.

Emerging Challenges in Tourism Destination Management in the Old Town

Chiang Mai's Old Town faces several challenges in sustaining its tourism industry while preserving its cultural and historical significance. Key concerns include heritage preservation, environmental sustainability, market shifts, foreign investment and seasonal air pollution. To remain competitive, Chiang Mai needs to develop strategies for proactive destination management, as sustainable tourism practices will be essential in shaping its future as a vibrant yet culturally preserved city.

Cultural Heritage Preservation and UNESCO Designation

One of the primary challenges is preserving Chiang Mai's Lanna heritage, which is central to its appeal for tourism. The city's traditional communities, language and cultural practices are increasingly at risk due to urbanisation and the influx of new residents. Social issues, including crime and environmental degradation, further threaten the city's heritage. Chiang Mai is currently pursuing UNESCO World Heritage status to mitigate these risks; however, this presents its own challenges. While the designation would enhance cultural preservation, it is likely to attract even greater numbers of tourists, potentially overwhelming local infrastructure and altering the character of the Old Town. Additionally, UNESCO status may impose restrictions on future development, limiting urban expansion and modernisation efforts (Betgem, 2022).

Air Pollution and Its Impact on Tourism

A significant environmental challenge for Chiang Mai's tourism sector is air pollution, particularly during Thailand's annual burning season (December to April). High levels of $PM_{2.5}$ pose serious health risks, discouraging outdoor activities and leading to a decline in visitor numbers. Tourism operators and industry leaders have raised concerns about the long-term impact on Chiang Mai's reputation, as the city is increasingly seen as a polluted destination (Beaulieu et al., 2023). Negative media coverage exacerbates this perception, although some reports also contain misinformation, such as claims regarding the closures of national parks. Managing Chiang Mai's image and mitigating the effects of pollution are essential for sustaining its appeal to tourists and safeguarding the quality of life for residents.

The annual burning season, driven by agricultural practices and forest fires, poses a persistent challenge for both residents and visitors. While some downplay its effects as part of Chiang Mai's climate cycle, tourism businesses report a noticeable decline in visitor numbers during these months. The presence of smog diminishes the city's visual appeal, impacts outdoor activities and negatively affects Chiang Mai's overall image as a tourist destination. Addressing environmental concerns will be crucial for ensuring the city's long-term sustainability as a preferred travel location.

Changing Tourist Behaviour and Market Shifts

While Chiang Mai remains a significant tourism destination, many visitors allocate limited time to exploring the Old Town itself. Instead, they prioritise activities outside the city, such as visiting elephant sanctuaries, national parks, waterfalls, ATV tours and zip-lining. Consequently, the Old Town often functions as a 'sleep city,' where tourists stay for accommodation and dining but engage in few immersive cultural experiences. Many visitors spend only one day exploring the temples and historic sites before moving on to other destinations.

Additionally, Western influences are reshaping the tourism landscape. Rapid globalisation and shifting tourist demographics are leading to an increase in Western-style accommodations, cafés and souvenir shops, weakening the city's authentic atmosphere. Traditional businesses are adjusting to meet the needs of foreign visitors, which, although economically advantageous, poses a risk to the erosion of local cultural identity.

Legalisation of Cannabis and the Rise of Cannabis Dispensaries in the Old Town

Following the post-COVID reopening of Thailand in 2022, the country decriminalised cannabis, primarily for medical and industrial use, but full recreational use also remains unregulated. As Chiang Mai recovers from the pandemic, cannabis legalisation remains a contentious issue. While it has created business opportunities and diversified tourism, it has also sparked debate among locals, authorities and business-owners. The rise of cannabis-related enterprises – dispensaries, cafés and wellness centres – has attracted both domestic and international visitors, boosting the hospitality and alternative medicine sectors in the Old Town.

However, the rapid, relatively unregulated expansion raises concerns about its long-term impact on Chiang Mai's cultural identity and family-friendly image. Initially promoted for medical and economic benefits, cannabis is now widely used recreationally, causing uncertainty as laws continue to evolve. The absence of clear regulations on consumption zones, licensing and enforcement creates confusion for tourists and businesses alike.

Public opinion remains divided, and law enforcement faces challenges in balancing regulation with business interests, whilst inconsistent policies fuel investor uncertainty. Although cannabis has economic potential, sustainable regulation and community engagement will ultimately determine whether it benefits or harms Chiang Mai's tourism industry in the long term.

Post-Pandemic Recovery and Declining Chinese Tourism

Tourism in Chiang Mai suffered significantly due to COVID-19, with visitor numbers dropping from 5.6 million pre-pandemic to 3.9 million in 2023 (Nation, 2024). Although recovery is underway, arrivals – particularly from China – remain lower than before, largely because of China's economic slowdown and stricter travel policies (Sukthaworn & Nuangjamnong, 2024). While the number of Chinese tourists has declined, an increasing number of Chinese expatriates are relocating to Chiang Mai, contributing to the local economy but also raising concerns over foreign property ownership.

European, Australian and North American markets, however, exhibit growth, with many visitors opting for hostels and guesthouses. Chinese tourists who visit Chiang Mai typically choose luxury hotels, while Western travellers prefer budget accommodations, highlighting a divergence in travel preferences.

Foreign Investment and Economic Leakage

Another critical issue is the rise in foreign investment, especially from Chinese investors purchasing properties in the Old Town. Many local stakeholders perceive this as a threat to Chiang Mai's cultural authenticity, as foreign-owned businesses cater primarily to their own markets, resulting in economic leakage – where profits remain within foreign communities instead of benefiting the local economy as a whole. This trend is particularly worrisome in the hospitality sector, where new luxury developments are transforming the city's historic centre.

Impact of Day Trips on Local Businesses

Chiang Mai's surrounding countryside, while a significant tourism asset, also presents challenges for the Old Town. Many visitors use Chiang Mai as a base for day trips to locations outside the city, such as Doi Inthanon, Pai and Chiang Rai, which leads to decreased local spending. Tour operators report that although Chiang Mai's landscape helps provide a diverse travel experience, the economic benefits frequently flow to outlying areas rather than businesses within the Old Town.

The Role of Tourism Stakeholders in Shaping the Future

The future of Old Town is not solely determined by government policies or tourism trends – it is shaped by multiple stakeholders with competing interests. Understanding the roles and influences of these stakeholders is essential for crafting a comprehensive and inclusive vision for the city's development.

Chiang Mai's Old Town is a vibrant, culturally rich and complex area, home to a diverse array of stakeholders who contribute to the tourism industry in various ways. These stakeholders comprise micro- and small-sized enterprises (MSEs), medium and large organisations (MLOs), government institutions, places of worship, international tourists and local communities. Understanding their roles, power dynamics and interactions is vital for sustainable tourism planning and policymaking. The current tourism stakeholders in the Old Town can be categorised as follows:

1. Micro- and Smallsized Enterprises (MSEs)

MSEs form the backbone of Chiang Mai's Old Town tourism sector, providing essential services such as accommodations, food, souvenirs and guided experiences. According to the Thai Ministerial Regulations on Small and Medium Enterprises Promotion (2019), micro-enterprises employ between 1 and 5 people, with an annual revenue not exceeding 1.8 million baht (equal to 53,000 USD), while small enterprises employ 6 to 30 people with an annual revenue of up to 300 million baht (equal to approximately 9 million USD). These businesses operate in several key tourism-related sectors:

- Souvenir shops and handicraft vendors – Selling traditional Thai crafts, textiles and handmade goods.
- Tourist information centres – Providing assistance, maps and tour bookings.
- Boutique hotels and guesthouses – Offering accommodation in both traditional and modern styles.
- Restaurants and street food stalls – Serving local and international cuisine.
- Cannabis dispensaries – Operating under Thailand's evolving cannabis regulations.

These enterprises make a substantial contribution to Chiang Mai's economy and cultural tourism experience, providing visitors with an authentic and personalised engagement with the city. However, many of them operate informally, which poses challenges for taxation and regulatory oversight (Çakmak, 2020). These businesses drive the city's tourism economy, but many struggle against corporate competition, rising rents and foreign investment takeovers (Choe & O'Regan, 2020).

2. Medium and Large Organisations (MLOs)
MLOs in Chiang Mai's Old Town comprise both public and private entities that maintain a more structured and long-term presence in the tourism sector. These organisations are classified as follows:

Public Institutions

- Government bodies – The Thai national and regional governments regulate tourism policies, infrastructure and development projects.
- Temples and religious institutions – Significant cultural sites that influence zoning laws and urban planning. Many temples charge a small entrance fee or rely on donations for their maintenance.
- Museums and historical sites – Offer cultural education and preserving Chiang Mai's heritage.
- Public educational institutions – Universities and vocational schools that contribute to tourism research and workforce training.

The Thai government's tourism development policies significantly influence Chiang Mai's growth trajectory. However, in previous decades, national tourism strategies have concentrated on Bangkok, Phuket and Pattaya, while Chiang Mai has received relatively less investment and attention (The Ministry of Tourism and Sports Thailand, 2017). Stronger regional policies are needed to ensure Chiang Mai's development aligns with sustainability goals and local priorities.

Private Organisations

- Hotels and hospitality businesses – While large international hotel chains face restrictions due to zoning laws, medium-sized boutique hotels, hostels and serviced apartments flourish.
- Restaurant chains and bars – Catering to tourists with a diverse range of culinary offerings, often concentrated in commercial areas.
- Retail and shopping centres – Including local handicraft markets, night markets and international franchises.
- Transportation services – Motorbike rentals, taxi services and travel agencies that facilitate mobility.
- Tour operators and travel agencies – Organising tours, excursions and experiences tailored for both international and domestic visitors.

Since the last decade, Chiang Mai has seen a rise in foreign investment, particularly from Chinese real estate developers and hospitality businesses (Siriphon, 2019). While such investments contribute to economic growth overall, they also raise concerns about economic leakage – where tourism revenues benefit foreign investors rather than local communities. Stronger regulations on foreign ownership and incentives for locally owned businesses could help address this issue. MLOs generally demonstrate more financial stability and regulatory compliance than MSEs. However, they also need to navigate cultural sensitivities, as seen in cases where businesses like nightclubs or establishments serving alcohol have struggled near religious sites due to community opposition.

3. Informal and Community Stakeholders

Beyond businesses and formal organisations, the Old Town of Chiang Mai is influenced by various community actors.

- Local residents – Their daily lives intersect with tourism, shaping policies and the viability of businesses.
- Domestic and international tourists – Their preferences and spending patterns influence market demand.
- Chinese investors and tour groups – Playing an increasingly significant role in Chiang Mai's tourism landscape through business ownership and organised travel packages.
- Non-profit organisations and cultural groups – Advocating for sustainable tourism practices and the preservation of heritage.

While tourism generates employment, it also reshapes the daily lives of residents. Numerous locals voice concerns regarding noise pollution, commercialisation and the erosion of traditional lifestyles (Betgem, 2022). Community-led initiatives, heritage conservation projects and policies that promote local participation in tourism planning are essential for maintaining Chiang Mai's cultural identity.

The interaction between formal enterprises, informal businesses and community groups contributes to the Old Town's unique character. However, this also creates challenges related to regulatory enforcement, economic competition and cultural preservation.

Power Dynamics Among Stakeholders

The distribution of power among stakeholders in Chiang Mai's Old Town is intricate and multifaceted. The Thai government possesses the highest level of authority, overseeing tourism policies, infrastructure development and regulations concerning foreign investment. For instance, zoning laws prohibit large hotel chains from operating within the Old Town, thereby preserving its historical character. Temples exert considerable influence beyond their religious role. Businesses that are not in harmony with Buddhist values, such as bars and nightclubs situated near temples, often face community opposition and struggle to survive.

Tourists possess considerable economic influence, as their expenditure underpins the prosperity of local businesses. Shifts in travel trends, such as an increased demand for sustainable tourism, can influence business priorities. Chinese investors are emerging as key stakeholders, with rising investments in local businesses, hotels and tour operations, raising concerns about economic dependence and cultural impact. Understanding these power relations is crucial for shaping Chiang Mai's tourism future.

Chiang Mai's Old Town is a rich tapestry of stakeholders, each playing a vital role in shaping its tourism landscape. From micro-businesses and informal vendors to temples, government agencies and international investors, the interactions between these groups define the city's identity. The coexistence of formal and informal economies, along with cultural and regulatory considerations, makes stakeholder management a crucial factor in ensuring the sustainable development of Chiang Mai as a premier tourist destination (Çakmak et al., 2018).

Preferable Future of Chiang Mai's Old Town: A Vision for Sustainable Growth and Cultural Preservation

The future of Chiang Mai's Old Town must be envisioned through a lens that harmonises economic growth with cultural preservation, ensuring that the area remains an attractive and sustainable destination for both tourists and residents. The northern and southern parts of the Old Town will develop along distinct, yet interconnected trajectories shaped by regional economic trends, evolving tourist behaviours and government policies. While the northern sector will prioritise cultural heritage and immersive local experiences, the southern district will emerge as a business and conference hub, reinforcing its role as an economic engine for the city. This bifurcation of development strategies acknowledges the diverse functions of Chiang Mai's historic urban core, ensuring that tourism expansion does not compromise the unique identity and heritage that define the city. In reaching the preferable future, the aspirations of tourism stakeholders play a vital role. These aspirations establish the connection between the preferred future and the interventions to be undertaken while considering the contextual issues for destinations and stakeholder groups.

Aspirations of Stakeholders in Chiang Mai's Old Town: Balancing Cultural Integrity and Economic Growth

Stakeholder aspirations in Chiang Mai's Old Town reflect a complex interplay between the preservation of cultural authenticity and the pressures of modernisation, economic sustainability and market-driven demands. Findings from stakeholder interviews suggest a broad consensus on the importance of maintaining Chiang Mai's distinctive cultural identity while recognising the need to adapt to shifting tourism patterns, infrastructural demands and evolving visitor expectations.

One of the most frequently expressed aspirations among local stakeholders is the preservation of Chiang Mai's cultural heritage, particularly in Old Town North, where traditional values and historical authenticity serve as key attractions. Local entrepreneurs, including café-owners, tour operators and street vendors, emphasise the importance of maintaining traditional Thai architectural styles, culinary customs and artisanal crafts. These aspirations are evident in initiatives such as the Sunday Night Market, which has evolved into a cultural staple that showcases authentic local products.

As one tour-company-owner observed:

> The market continues to be a space where local people and handmade souvenirs take precedence over mass commercialisation.
>
> (Interview 17, 2024)

Nevertheless, there is concern that other aspects of local culture are not adequately promoted, leading tourists to prefer newly established Western-style cafés and restaurants instead of engaging with traditional local businesses. This observation aligns with Cohen (1988)'s findings that over-commercialisation of tourism can undermine local identity and divert consumer preferences away from traditional option offerings.

While preserving cultural heritage remains a priority, there is also recognition of the necessity for modernisation to meet evolving visitor expectations. The tourism sector in Chiang Mai faces a growing demand for enhanced infrastructure, improved public transport and contemporary hospitality services. Business-owners are keen to renovate and modernise their establishments to attract younger, tech-savvy visitors, particularly digital nomads and long-term expatriates. As one café-owner remarked:

> I want to renovate my business in the future and make it more modern and appealing to digital nomads.
>
> (Interview 9, 2024)

These aspirations underscore a wider trend in the tourism industry, where businesses must reconcile heritage preservation with modernisation, ensuring that while facilities enhance, they do not compromise the city's cultural essence (Su, Bramwell & Whalley, 2018).

Another significant theme emerging from stakeholder interviews is the uneven distribution of tourism revenue and engagement within the city. Many stakeholders express frustration that, although Chiang Mai boasts a diverse range of attractions, tourists tend to spend the majority of their time and money outside the Old Town, participating in adventure tourism activities such as ATV rides, elephant sanctuaries, rafting and zip-lining (Interview 6, 2024). When visitors do choose to remain in the Old Town, they usually restrict their experiences to a limited number of well-known temples and food markets, as noted by a hostel operator:

> Tourists only visit the temples and some food markets.
>
> (Interview 4, 2024)

This concentrated flow of tourists has economic implications, as local businesses outside of these main attractions struggle to benefit from tourism revenue. The lack of diversified experiences in the Old Town exacerbates concerns that Chiang Mai is becoming a 'sleep town,' where tourists stay overnight but engage in activities elsewhere (Interview 29, 2024).

One of the most pressing concerns among stakeholders is the gradual shift in Chiang Mai's commercial landscape, which has resulted in the diminishing visibility of local culture within the tourism sector. As international visitors, particularly digital nomads, gravitate towards modern cafés, co-working spaces and Western-style restaurants, many local business-owners feel compelled to modify their offerings to stay competitive. A local restaurant-owner conveyed their concern over this trend, stating:

> I am missing local food because more restaurants with Western food are coming.
>
> (Interview 11, 2024)

Another stakeholder noted that digital nomads and younger tourists favour 'modern places,' prompting restaurants in Old Town to adapt their aesthetics and offerings (Interview 10, 2024). While this transformation may enhance Chiang Mai's appeal to new visitor demographics, it also raises concerns about cultural dilution and the erosion of local identity (Gotham, 2005).

Another prominent challenge highlighted in stakeholder interviews is the inadequacy of Chiang Mai's tourism promotion strategies. Stakeholders frequently voice dissatisfaction with the lack of effective marketing campaigns, both domestically and internationally. A hostel manager remarked that:

> Chiang Mai lacks promotion; the tourism board doesn't know how to put Chiang Mai in the international tourism markets.
>
> (Interview 16, 2024)

While another interviewee noted that when the city is promoted, the content is 'often in Thai' (Interview 16, 2024), which limits its accessibility to foreign audiences. Furthermore, although governmental support for tourism development has traditionally played a significant role in attracting visitors, some stakeholders argue that current efforts are inadequate, asserting:

> Normally, the government does assist us in attracting tourists. This year, they claim to help, but we do not perceive this.
>
> (Interview 7, 2024)

These concerns underscore the need for more targeted and multilingual marketing efforts to engage broader audiences and ensure Chiang Mai remains a competitive international destination (Pike & Page, 2014).

Stakeholder aspirations also highlight the necessity of broadening Chiang Mai's market segmentation, aiming to attract a more diverse tourist base. The city's tourism economy has traditionally depended heavily on European backpackers and long-term travellers, as noted by one hostel receptionist:

> Almost all the tourists come from Europe.
>
> (Interview 11, 2024)

However, Chiang Mai is also witnessing increasing demand from Asian markets, particularly from Chinese, South Korean and Indian tourists, who often seek luxury accommodations and higher-end experiences. This distinction is vital for future tourism development, as it underscores the potential for segment-specific tourism strategies that cater to both budget-conscious travellers and high-spending markets. Destinations that effectively differentiate their offerings for various traveller segments can maximise their economic benefits while ensuring long-term resilience.

In addition to these broader tourism-related aspirations, Chiang Mai's MICE sector is undergoing a significant transformation, particularly in the upscaling of services and facilities. A key aspiration among larger businesses and tourism stakeholders is to enhance staff competencies, language proficiency and customer service quality to attract high-value business travellers. A representative from the Thailand Convention and Exhibition Bureau (TCEB) remarked that improved service quality is a core component of MICE tourism:

> Tourists will be able to better understand the vendors; they will communicate more effectively in the area of the night markets.
>
> (TCEB Interview, 2024)

Simultaneously, extensive redevelopment projects, like Asset World Corporation's (AWC) revitalisation of the Night Bazaar, are continually transforming the city's commercial districts to serve corporate travellers and affluent clientele. The aspirations of Chiang Mai's diverse stakeholders reflect a shared desire to balance cultural preservation with economic modernisation. While local entrepreneurs, tour operators and residents stress the necessity of safeguarding traditional crafts, cuisine and historic sites, businesses catering to international visitors recognise the significance of modernised services and infrastructure. The interviews highlight a growing divide between cultural conservation and commercial adaptation, emphasising the challenges presented by tourism concentration, shifting consumer preferences and ineffective marketing strategies. As Chiang Mai evolves, addressing these aspirations through collaborative policy frameworks, sustainable tourism initiatives and diversified marketing efforts will be crucial in preserving its unique identity while reinforcing its status as a leading tourism destination.

Preferable Future Vision for Southern and Northern Parts of the Old Town

Northern Part

In the northern part of the Old Town, the preferable future is one where tourism nurtures a vibrant, lived-in environment rather than a transient space where visitors merely stay overnight before seeking experiences elsewhere. The challenge of preventing the area from devolving into a 'sleep city' will necessitate focused efforts to create engaging attractions that encourage tourists to prolong their visits and actively engage with local culture. Central to this transformation is the enhancement of iconic sites through the development of compelling narratives that connect visitors to the city's rich history. By incorporating storytelling into tourism experiences, attractions such as ancient temples, historical alleyways and traditional markets will be portrayed not just as static sites but as living testimonies to Chiang Mai's cultural evolution. This narrative-driven approach will enable tourists to experience the city through the eyes of its residents, fostering a deeper appreciation for its traditions and way of life.

The rise of digital tourism has significantly changed how travellers discover and experience destinations, making online visibility a crucial factor in shaping visitor behaviour. The growing reliance on social media platforms such as Instagram, Facebook, TikTok and travel blogs has led to a phenomenon where tourists tend to follow prescribed routes based on popular online recommendations. While this has undoubtedly enhanced Chiang Mai's appeal, it has also posed challenges in maintaining the authenticity of the Old Town's cultural landscape. In the preferable future, digital strategies will be utilised not merely to attract more tourists but to guide them towards experiences that align with the city's cultural identity. The implementation of curated promotional campaigns that highlight lesser-known attractions – such as hidden temples, artisan workshops and local culinary gems – will diversify tourist activity, alleviating over-tourism in heavily trafficked areas while ensuring that smaller businesses benefit from visitors' spending.

The emphasis on local authenticity as a core element of the northern Old Town's tourism strategy will be reinforced through initiatives such as the development of a 'Hidden Gems Map,' which will serve as a guide to culturally significant yet underappreciated locations. This initiative will complement broader efforts to decentralise tourism within the city, ensuring that the economic benefits of increased visitation are distributed more evenly. Furthermore, by integrating story-telling into tourism products, visitors will be encouraged to engage with the city beyond its surface-level attractions, participating in activities that connect them with Chiang Mai's living heritage. These experiences may include guided historical walks, immersive art installations that narrate the city's past and interactive culinary tours that introduce visitors to the intricacies of northern Thai cuisine. By positioning Chiang Mai as a destination that offers not only sightseeing opportunities but also meaningful cultural exchanges, the northern Old Town will cultivate a tourism model that is both sustainable and enriching.

Southern Part

At the same time, the southern part of the Old Town is set to undergo a transformation into Chiang Mai's primary hub for business and conference tourism. With the ongoing expansion of the MICE sector in Thailand, Old Town South will become a focal point for corporate travel, attracting business visitors from across Southeast Asia and beyond. This development will align with broader regional economic trends, as the rising affluence of Southeast Asian economies is anticipated to drive greater demand for high-quality business tourism experiences. The presence of well-established corporate hotels and event venues in this district provides a robust foundation for such growth, and future investments will further enhance the area's capacity to host large-scale conferences and professional gatherings.

The economic benefits of strengthening the MICE sector in Chiang Mai will extend beyond the corporate realm, influencing the broader tourism landscape of the city. As business travellers seek opportunities to combine work with leisure – commonly referred to as 'bleisure' travel – there will be an increasing demand for high-end tourism experiences that integrate cultural and recreational elements. This will create opportunities for local entrepreneurs to develop tourism products tailored to business visitors, such as exclusive culinary experiences, curated cultural tours and luxury wellness retreats. The infusion of MICE tourism into the city's economic fabric will not only generate revenue for local businesses but will also enhance Chiang Mai's reputation as a destination that seamlessly blends professional engagement with cultural immersion.

In less than a decade, the cumulative outcomes of these strategic developments could elevate Old Town South to become one of Thailand's leading MICE destinations. The Night Bazaar area, in particular, will play a crucial role in facilitating this transformation, serving as both an economic centre and a vibrant social space where business travellers and leisure tourists converge. The anticipated rise in conference tourism will drive increased demand for luxury accommodations, prompting corporate hotels to focus more on serving the business travel market. This shift will, in turn, influence the broader accommodation landscape, as budget-conscious leisure travellers seek alternative lodging options within the Old Town walls. The resulting diversification in the hospitality sector will ensure that Chiang

Mai remains accessible to a wide range of visitors while maximising economic gains from high-spending business tourists.

Despite the anticipated economic growth in both northern and southern Old Town, challenges associated with urbanisation, environmental sustainability and cultural preservation will require ongoing attention. The growing presence of foreign investment – particularly from Chinese stakeholders – has already started to reshape the city's commercial landscape, raising concerns about the long-term effects of globalisation on local businesses and traditions. The relationship between foreign and local enterprises will need to be carefully managed to prevent the displacement of small businesses and the erosion of Chiang Mai's unique character. Likewise, the continued expansion of tourism will demand infrastructural improvements, especially in transportation and waste management, to ensure that the rise in visitor numbers does not compromise the quality of life for residents.

The government's role in shaping the future of Old Town will be critical, as regulatory frameworks must be adapted to balance economic incentives with cultural preservation. While zoning laws currently restrict the construction of large-scale hotels within the Old Town, ongoing monitoring will be necessary to assess whether additional measures are needed to protect the historic environment. At the same time, policies that support local entrepreneurship – such as grants for small businesses and incentives for community-based tourism initiatives – will be crucial in maintaining the economic vitality of the region.

Ultimately, the preferable future for Chiang Mai's Old Town is one that embraces change while preserving the cultural heritage that defines the city. The northern district will strengthen its reputation as an authentic and immersive cultural destination, while the southern sector will develop into a thriving business hub that blends corporate tourism with local traditions. Through strategic planning, responsible governance and community involvement, Chiang Mai can ensure that its Old Town remains a place where history and modernity coexist harmoniously, providing economic prosperity without compromising the essence of the city.

Recommendations and Implementable Interventions for Sustainable Tourism Development in Chiang Mai's Old Town

The sustainable and culturally sensitive development of Chiang Mai's Old Town necessitates well-structured interventions that align with the unique character of the destination while ensuring economic and social benefits for local stakeholders. Based on an analysis of the current tourism landscape and future projections, three feasible interventions are proposed: (1) the establishment of a community-led brand ambassador programme, (2) the development of a Hidden Gem Map to decentralise tourism flow and (3) the introduction of a guided street art tour integrating cultural narratives. These interventions are designed to enhance tourist engagement, preserve the authenticity of the destination and support local businesses in a sustainable manner.

Establishment of a Community-Led Brand Ambassador Programme

To foster a deeper connection between tourists and the local culture, it is recommended to introduce a brand ambassador programme, comprising local residents who serve as representatives of Chiang Mai's Old Town North. This initiative draws upon the concept of community-based tourism (CBT), which highlights active local participation in shaping the visitor experience (Scheyvens, 2002). The role of these ambassadors would extend beyond conventional tour guiding to include cultural storytelling, historical interpretation and the promotion of sustainable tourism practices.

Brand ambassadors should be chosen from knowledgeable and passionate locals, including long-term residents, historians, artisans and business-owners who possess a strong connection to Chiang Mai's heritage (Rehmet & Dinnie, 2013). Their primary responsibility would be to communicate the authentic essence of the Old Town, ensuring that tourism development aligns with cultural preservation. These ambassadors could conduct workshops, share historical narratives and create digital content that highlights lesser-known aspects of the city. Their engagement would occur both offline – through guided walks, storytelling sessions and community events – and online, via social media platforms where they can share narratives with a wider audience.

The programme's success would depend on collaborative governance, necessitating participation from key stakeholders such as municipal tourism authorities, local businesses and educational institutions. The government's role would involve providing logistical and financial support, ensuring that brand ambassadors receive proper training in storytelling techniques, cross-cultural communication and sustainable tourism principles. Local businesses, in turn, would benefit by forming partnerships with ambassadors to present their offerings in an authentic and contextually appropriate manner.

This initiative is anticipated to yield several advantages. First, it would heighten visitor engagement as tourists seek interactive and immersive experiences that extend beyond traditional sightseeing. Second, bolster cultural preservation, ensuring that narratives passed down through generations remain central to the tourist experience. Lastly, it could amplify word-of-mouth promotion, as visitors exposed to authentic local stories are more inclined to share their experiences, thereby enhancing Chiang Mai's global reputation as a culturally rich destination.

Development of a Hidden Gem Map to Promote Equitable Tourism Distribution

A second critical intervention would be the creation of a Hidden Gem Map, designed to diversify tourist activity and distribute visitor flow more evenly across Old Town North. Currently, the concentration of tourism in a few well-known areas has led to overcrowding at popular sites while leaving many culturally significant locations underappreciated. The map would highlight lesser-known temples, artisanal workshops, family-run restaurants and historical alleyways, offering tourists a curated selection of experiences that are deeply rooted in local traditions.

The map would be produced in both digital and physical formats, ensuring accessibility for different types of travellers. A paper version could be distributed at hotels, tourist information centres and Chiang Mai International Airport, while an interactive digital version could be made available through a dedicated website and QR codes linked to social media pages. The digital component would also feature short documentary-style videos and ambassador-led narratives, allowing tourists to gain deeper insights into each location before visiting.

Beyond merely guiding tourists to new locations, the Hidden Gem Map would serve an educational purpose by embedding local narratives and cultural explanations within each featured site. Rather than simply listing points of interest, the map could be organised around thematic routes, such as:

- 'Sacred Spaces and Spiritual Stories' – Highlighting small, lesser-known temples with unique histories.
- 'Flavours of the Old Town' – Showcasing family-owned restaurants that specialise in traditional Lanna cuisine.
- 'Artisan Trails' – Directing visitors to workshops where local craftsmen create textiles, wood carvings and pottery.
- 'Forgotten Alleys and Secret Gardens' – Introducing visitors to hidden courtyards, murals and quiet historical spaces' significance.

To ensure its credibility and effectiveness, the map should be co-developed with local businesses and cultural institutions, guaranteeing that all featured locations align with the objectives of sustainable tourism and local economic development. Tourist engagement could be monitored through digital interactions, such as the number of QR code scans and website visits, providing measurable indicators of success.

This intervention would have several positive impacts. It could encourage longer stays as tourists explore new areas and experiences that extend their itineraries within the Old Town. It would also support local businesses by directing visitor spending towards smaller enterprises that might otherwise be overshadowed by larger, more commercial establishments. Lastly, it could alleviate overcrowding in heavily trafficked areas, thereby minimising the negative environmental and social effects of concentrated tourism.

Introduction of a Guided Street Art Tour Integrating Cultural Narratives

The third proposed intervention involves developing a guided street art tour that would highlight Chiang Mai's evolving artistic landscape while integrating historical and cultural narratives into the experience. Although internationally recognised for its temples and markets, Chiang Mai's Street Art scene remains an underutilised asset in tourism development. This initiative aims to reposition urban art as a key cultural attraction, utilising it to connect contemporary artistic expression with the city's historical identity.

The street art tour should be created in collaboration with local artists, cultural historians and tour guides, ensuring that each mural is placed within its broader socio-cultural context. Visitors would not only admire the visual artistry but also gain insights into the symbolism and stories behind each piece. Some murals, for instance, depict Buddhist teachings, while others reflect Lanna history, social movements or contemporary issues such as environmental conservation. By connecting these artistic expressions to Chiang Mai's heritage, the tour would transform street art from mere decoration into a narrative medium that enhances visitor understanding.

The tour should be structured as a hybrid experience, combining guided walks with self-guided options. Whilst professional tour guides could offer scheduled group tours, independent travellers would have the opportunity to explore at their own pace using an interactive map with embedded audio narratives, accessible via QR codes placed near key locations' murals.

To maximise its impact, the street art tour should be integrated into broader tourism promotions, appearing on the Hidden Gem Map and being featured in social media campaigns led by the brand ambassador programme. Furthermore, partnerships with local cafés and restaurants would facilitate the incorporation of authentic culinary experiences within the tour, enhancing its appeal to food-oriented travellers.

This intervention is anticipated to yield numerous benefits, enhancing visitor engagement with contemporary Thai culture, cultivating a deeper appreciation of Chiang Mai that goes beyond its historical sites, and generating new revenue opportunities for local artists by offering chances for visitors to purchase art prints or commission bespoke pieces. Additionally, it could bolster urban regeneration efforts, highlighting underutilised spaces that can be revitalised through public art initiatives.

These proposed interventions – a brand ambassador programme, a Hidden Gem Map and a guided street art tour – represent feasible, contextually appropriate solutions for achieving the preferable future of Chiang Mai's Old Town. Each initiative is grounded in principles of sustainable tourism, cultural preservation and economic inclusivity, ensuring that the growth of the tourism sector benefits local communities while enhancing visitor experiences. By fostering deeper engagement, promoting decentralised tourism and integrating modern storytelling techniques, Chiang Mai can position itself as a model destination that successfully balances heritage conservation with innovative tourism development. These recommendations help provide a roadmap for the city to evolve in a manner that respects its past while embracing its future, ensuring that the Old Town remains a vibrant, living heritage site for generations to come.

Conclusion

Chiang Mai's Old Town stands at a crossroads where the pressures of modernisation, shifting tourist demographics and economic demands intersect with a strong desire to preserve its cultural authenticity. Tourism remains the economic backbone of the Old Town; however, stakeholders – including local businesses, government authorities, cultural institutions and residents – confront the complex challenge of sustaining growth while upholding the city's historical and cultural integrity. The findings from stakeholder interviews elucidate key aspirations, challenges and potential strategies to shape a preferable future. Examining these elements in depth offers insight into the necessary interventions required to ensure that Chiang Mai's Old Town remains a thriving yet culturally preserved destination.

The primary challenge facing Chiang Mai's Old Town is the gradual erosion of cultural authenticity due to commercial pressures, changing consumer preferences and the increasing presence of modern businesses catering to international tourists. Many stakeholders, particularly local business-owners, have expressed concerns that while Chiang Mai is still viewed as an authentic Thai cultural destination, the growing number of Western-style cafés, restaurants and co-working spaces has diluted its traditional identity. Some interviewees noted a decline in local cuisine and cultural representation as businesses adapt to meet the needs of digital nomads and younger tourists seeking contemporary environments. This trend aligns with broader research on tourism-driven gentrification, which highlights how unregulated tourism growth can lead to the displacement of traditional businesses and the homogenisation of local culture. As Chiang Mai continues to attract a diverse array of visitors, it faces the risk of losing elements of its unique character, a concern shared by both local entrepreneurs and long-term residents.

Another significant challenge is the limited distribution of tourism revenue within the Old Town. While Chiang Mai offers a diverse array of cultural and experiential attractions, most tourists engage in activities located outside the Old Town, participating in adventure tourism and visits to elephant sanctuaries. Within the Old Town itself, visitor engagement is predominantly confined to major temples and food markets, leaving other areas economically disadvantaged. The concentration of tourism in specific areas exacerbates income inequality, with smaller businesses and lesser-known cultural sites struggling to attract visitors and generate revenue. The absence of incentives to encourage tourists to explore beyond the primary attractions further contributes to an economic imbalance, making it challenging for local entrepreneurs to compete with more commercially successful establishments.

A further challenge lies in the ineffective and poorly executed tourism marketing and promotional strategies in Chiang Mai, which fail to attract and engage visitors effectively. Stakeholders frequently express dissatisfaction with the lack of effective outreach to international audiences, noting that existing promotional materials are often in Thai and fail to highlight the city's diverse offerings. Many business-owners and tourism professionals feel that Chiang Mai has not effectively positioned itself in the global tourism market, leading to missed opportunities for attracting a broader and more diverse array of visitors. Inadequate digital marketing efforts, limited multilingual content and a failure to differentiate Chiang Mai from competing Southeast Asian destinations have further hindered its potential for sustainable tourism growth. The absence of targeted branding strategies has resulted in a reliance on organic word-of-mouth promotion, which, while beneficial in some cases, is insufficient for securing long-term tourism development. Without a coordinated and strategic marketing approach, Chiang Mai risks losing its competitive edge to other destinations that are more actively engaged in promoting their unique cultural and economic assets.

In addition to these broader tourism-related challenges, Chiang Mai's MICE sector is undergoing a significant transformation, presenting both opportunities and constraints. While the sector has the potential to drive high-value tourism and contribute to the Old Town's economic resilience, it requires substantial investments in infrastructure, service quality and professional training. As international business travellers expect a higher standard of service, there is a growing emphasis on improving staff competencies, language proficiency and overall customer experience. Large-scale redevelopment projects, such as the planned revitalisation of the Night Bazaar, aim to modernise Old Town's commercial districts and make them more appealing to corporate visitors. However, ensuring that such projects benefit local communities while maintaining cultural authenticity remains a key concern. If not managed carefully, the expansion of the MICE sector may contribute to further commercialisation and the displacement of traditional businesses, exacerbating the existing tensions between economic development and cultural preservation.

Stakeholder aspirations reveal a collective desire for a tourism model that balances cultural preservation with economic sustainability. Business-owners, artisans and local residents share a strong commitment to maintaining Old Town's cultural identity, ensuring that local traditions, cuisine and craftsmanship remain integral to the visitor experience. At the same time, there is recognition of the need for modernisation and infrastructure improvements, particularly in areas such as public transport, digital marketing and hospitality services. The ideal future for Chiang Mai's Old Town is one where cultural heritage and economic diversification are not mutually exclusive but rather are complementary forces contributing to the city's long-term resilience. Instead of relying on mass tourism and generic experiences, the Old Town should position itself as a hub for immersive cultural engagement, where visitors actively participate in local traditions, storytelling and artistic expression. Decentralising tourism activities will ensure that lesser-known attractions and businesses benefit from increased foot traffic, reducing economic disparities within the city and promoting a more inclusive tourism economy. Additionally, the Old Town's MICE sector will need to develop in a manner that enhances service quality while preserving Chiang Mai's historic and cultural charm.

To achieve the preferable future, a multi-stakeholder approach is necessary, integrating strategic interventions that support cultural preservation, economic diversification and infrastructural enhancement. In addition, a comprehensive digital marketing strategy is vital for enhancing Old Town's, and generally Chiang Mai's, international positioning. This approach should utilise social media, multilingual content and influencer collaborations to engage international travellers more effectively. A dedicated multilingual tourism website that features immersive video content, traveller testimonials and curated itineraries would offer a central platform for prospective visitors to interact with the city prior to their arrival. By adopting a more proactive marketing strategy, Old Town can attract a wider range of tourists and reduce its current dependence on organic visitor flows.

Finally, as the MICE sector continues to develop, investments in staff training, infrastructure and service quality will be vital. Hospitality professionals should receive ongoing training in foreign language skills, cross-cultural service delivery and digital marketing to ensure Chiang Mai remains competitive in the business tourism market. Efforts to develop 'bleisure' experiences, which blend business and leisure activities, will further enhance the city's appeal to corporate travellers, while ensuring that the economic benefits of MICE tourism extend beyond conference centres and hotels.

The future of Chiang Mai's Old Town hinges on finding a balance between tradition and innovation. By implementing strategic, community-led interventions, the city can maintain economic growth while safeguarding its unique cultural heritage. Collaborative efforts among government agencies, local businesses and community representatives will be crucial in shaping a sustainable and inclusive tourism model that ensures Chiang Mai remains a globally renowned cultural and business tourism destination.

References

Beaulieu, A., Leblond, J. P., Déry, S., & Cao, H. (2023). Urban air pollution anxieties, forest conservation, and farmland expropriation: State territorialization in the uplands and highlands of northern Thailand. *Land Use Policy*, *131*, 106687.

Betgem, F. (2022). Chiang Mai UNESCO World Heritage Site – Chiang Mai à La Carte. Chiang Mai À La Carte. Available at www.chiangmai-alacarte.com/chiang-mai-unesco-world-heritagesite/

Çakmak, E. (2020). *The practice of informal tourism entrepreneurs: A Bourdieusian perspective* (Doctoral dissertation, Wageningen University and Research).

Çakmak, E., Lie, R., & McCabe, S. (2018). Reframing informal tourism entrepreneurial practices: Capital and field relations structuring the informal tourism economy of Chiang Mai. *Annals of Tourism Research*, *72*, 37–47.

Çakmak, E., Tucker, H., & Hollinshead, K. (2021). Introduction – Tourism paradoxes: Contradictions, controversies and challenges. In E. Çakmak, H. Tucker & K. Hollinshead (eds.), *Tourism Paradoxes* (pp. 1–14). Bristol: Channel View.

Choe, J., & O'Regan, M. (2020). Faith manifest: Spiritual and mindfulness tourism in Chiang Mai, Thailand. *Religions*, *11*(4), 177.

Cohen, E. (1988). Authenticity and commoditization in tourism. *Annals of Tourism Research*, *15*(3), 371–386.

Gotham, K. F. (2005). Theorizing urban spectacles. *City*, *9*(2), 225–246.

Mostafanezhad, M., & Promburom, T. (2018). 'Lost in Thailand': The popular geopolitics of film-induced tourism in northern Thailand. *Social & Cultural Geography*, *19*(1), 81–101.

Nation (2024). Available at www.nationthailand.com/thailand/economy/40034360 (Accessed 2 March, 2025).

Pike, S., & Page, S. J. (2014). Destination Marketing Organizations and destination marketing: A narrative analysis of the literature. *Tourism management*, *41*, 202–227.

Ploysri, P. (2023). Tourism development in Chiang Mai: Growth and diversity. Personal Communication at the Chiang Mai University, Thailand.

Rehmet, J., & Dinnie, K. (2013). Citizen brand ambassadors: Motivations and perceived effects. *Journal of Destination Marketing & Management*, *2*(1), 31–38.

Scheyvens, R. (2002). *Tourism for Development: Empowering Communities*. England, Pearson Education.

Siriphon, A. (2019). Developing entrepreneurship under the rise of China: Chinese migrant entrepreneurs in tourism-related businesses in Chiang Mai. *The Sociology of Chinese Capitalism in Southeast Asia: Challenges and Prospects*, 271–289.

Statista. (2024). Available at www.statista.com/chart/32594/projected-start-of-population-decline/ (Accessed 22 February 2025).

Su, R., Bramwell, B., & Whalley, P. A. (2018). Cultural political economy and urban heritage tourism. *Annals of Tourism Research*, *68*, 30–40.

Sukthaworn, M., & Nuangjamnong, C. (2024). Influencing factor of price, facilities, and destination image on customer satisfaction toward glamping revisit intention in Chiang Mai, Thailand. In *AU Hybrid International Conference 2024 on "Entrepreneurship & Sustainability in the Digital Era" under the theme of "People Centric Knowledge in Intelligence World"* University of Thailand, (Vol. 4, No. 1, pp. 508–533).

The Ministry of Tourism and Sports Thailand. (2017). The Second National Tourism Development Plan (2017–2021). Office of Permanent Secretary Ministry of Tourism and Sports. Roland Berger. Available at `www.google.com/url?sa=t&rct=j&q=&esrc=s&source=web&cd=&ved=2ahUKEwj9w6f3zt39AhUjDbcAHRjACqYQFnoECBAQAQ&url=https%3A%2F%2Fwww.rolandberger.com%2Fpublications%2Fpublication_pdf%2Froland_berger_the_second_national_tourism_development_plan_2017_2021.pdf&usg=AOvVaw0Vzx4-mvzFmObNYOtK_mPC

Trading Economics. (2025). Available at https://tradingeconomics.com/thailand/gdp (Accessed 22 February 2025).

UNESCO World Heritage Centre. (2015). Monuments, sites and cultural landscape of Chiang Mai, Capital of Lanna – UNESCO World Heritage Centre. Available at https://whc.unesco.org/en/tentativelists/6003/

UNESCO (2024) UNESCO heritage programme Penang. Available at https://unesco.org.my/v2/unesco-heritage-programme-penang-island/ (Accessed 4th May, 2024)

Visit Chiang Mai. (2020). A short history of the Lanna Kingdom around Chiang Mai. Available at www.visitchiangmai.com.au/history.html

World Bank. (2024). Available at www.worldbank.org/en/country/thailand/overview (Accessed 22 February 2025).

9 The Preferable Future of the Nimman Area, Chiang Mai, Thailand

Introduction

Nimmanhaemin, commonly referred to as Nimman, has undergone a profound transformation over the past few decades, evolving from a planned residential neighbourhood into a major commercial and tourism hub within Chiang Mai. Initially designed by the Nimmanhaemin family in the 1960s as a residential area, its trajectory shifted in the late 1990s with the emergence of commercial enterprises. A significant turning point in this transition was the establishment of the Rincom Hotel at the intersection of Huay Kaew Road and Nimmanhaemin Road, which catalysed increased visitor traffic and business activity. The subsequent rise of creative industries and independent entrepreneurs, supported by events such as the Nimmanhaemin Art and Design Promenade (NAP), further contributed to the area's development into a centre for cultural production, innovation and tourism. Today, Nimman is widely recognised for its fusion of modern urban amenities and creative enterprises, distinguishing itself within the broader landscape of Chiang Mai's urban transformation.

In recent years, Nimman has emerged as a key destination for digital nomads, young professionals and expatriates. Chiang Mai is consistently ranked among the top global locations for remote work, with Nimman serving as the epicentre of this trend. According to Flatio (2024), Chiang Mai ranks among the top-ten most popular cities for digital nomads globally, with only Bangkok surpassing it in Thailand. The area's appeal is further reinforced by perceptions of safety and affordability, particularly among female travellers seeking long-term stays. The integration of co-working spaces, boutique accommodations and international cafés has contributed to the district's reputation as an ideal environment for remote professionals. Additionally, the architectural and visual aesthetics of Nimman play a central role in its tourism appeal. The prevalence of stylish cafés, boutique hotels and contemporary retail spaces has established Nimman as a visually appealing 'Instagrammable' destination. Participant observations of visitor behaviours in the area and online media analysis indicate a strong emphasis on photography and content creation, reflecting the influence of digital media in shaping the district's cultural and commercial identity. The commercial core of Nimman, particularly

DOI: 10.4324/9781003570240-12

around Maya Lifestyle Shopping Center and One Nimman, epitomises this aesthetic orientation, combining international retail brands with design-centric urban spaces that appeal to a globalised consumer base.

The economic landscape of Nimman has also been shaped by broader urban development strategies, particularly Chiang Mai's positioning as a SMART city. Since 2018, the district has been a focal point for infrastructure investment, digital innovation and real estate development (Thai PBS, 2019). These initiatives have accelerated the conversion of residential properties into commercial spaces, aligning with broader trends of urban modernisation and tourism-driven economic growth. The relative affordability of living and operating businesses in Nimman compared with Western cities has further incentivised foreign investment and long-term expatriate residency. However, this rapid transformation has led to significant changes in land use patterns, raising questions about the sustainability of growth and the preservation of local identity.

Despite its success as a commercial and tourism hub, Nimman faces several challenges that warrant critical examination. Over-tourism has emerged as a pressing concern, particularly following the surge of Chinese visitors after the release of the 2012 film *Lost in Thailand* (Mostafanezhad & Promburom, 2018). The influx of tourists has generated tensions between economic development and local interests, exemplified by controversies such as the operation of 'tourist road trains' through Chiang Mai University (CMU), which has been met with resistance from students who argue that it commodifies their educational environment. In addition, Nimman's commercial landscape is characterised by volatility, with small enterprises frequently adjusting their business models in response to shifting consumer trends. The COVID-19 pandemic further exposed the vulnerabilities of this economic model, as travel restrictions led to widespread business closures and financial instability. While tourism has since rebounded, the district's heavy dependence on international visitors, particularly from East Asia, underscores the need for a more diversified and resilient economic framework.

This chapter examines Nimman's preferable future trajectories in light of ongoing economic, technological and urban transformations. As the district continues to evolve, it is imperative to consider strategies that balance economic expansion with cultural preservation and community well-being. The challenge lies in sustaining Nimman's role as a dynamic and globally connected district while ensuring that development remains inclusive and responsive to local needs. By exploring various scenarios and strategic possibilities, this chapter seeks to contribute to a broader understanding of how Nimman can navigate the complexities of tourism, urbanisation and economic change within a rapidly changing Asian context.

Methodology

This study employs a qualitative research approach. Given the complexity of the area's socio-economic dynamics, a multi-method strategy was adopted, incorporating field observations, stakeholder interviews media and document analysis. This triangulated approach enhances the reliability of findings by capturing multiple perspectives on the district's commercial landscape, tourism trends, infrastructure challenges and stakeholder expectations.

Primary data collection involved systematic field observations and semi-structured interviews with key stakeholders. Observational research focused on mapping the district's spatial and business composition, with particular attention to commercial hotspots such as Maya Lifestyle Shopping Center, One Nimman and surrounding streets. Data collection included assessments of business typologies, visitor demographics, traffic congestion, pedestrian accessibility and seasonal fluctuations in commercial activity. These observations provided a foundational understanding of the district's urban form and economic functionality.

To gain direct insights from key actors, 42 stakeholder interviews were conducted with business owners, employees, residents, investors, public officials and tourists. Interviews followed a structured framework to ensure consistency while allowing for open-ended discussion. In addition, two expert interviews were conducted with social sciences scholars from Chiang Mai University. Following these expert interviews, 11 informal conversations with employees from hotels, cafés, restaurants and service-based businesses captured perspectives on market dynamics, operational challenges and evolving consumer behaviour. Additionally, four interviews were conducted with digital nomads to understand their locational choices, working conditions and long-term expectations for the district. When a local language barrier arose, a translator was used where necessary to ensure accurate data collection. To facilitate comparative analysis, participants were asked to describe their vision for Nimman's future in a single word.

In addition, online media research was undertaken to analyse online discourse surrounding Nimman's development. Social media platforms such as Instagram, TikTok and Facebook, as well as travel blogs and digital nomad forums were examined to identify recurring themes in visitor experiences, consumer preferences and tourism marketing strategies. Engagement with online communities further enriched qualitative findings; when a Facebook group for expatriates and locals in Chiang Mai was queried about Nimman's future, eight respondents provided detailed perspectives.

To complement primary research, secondary data sources were reviewed, including government reports, academic literature and industry analyses. Data from institutions such as the Thailand Board of Investment (BOI), the Tourism Authority of Thailand (TAT) and local economic bodies provided macroeconomic and social contexts regarding tourism flows, real estate trends and investment patterns. Additionally, digital platforms such as Nomad List and Flatio were analysed to assess Chiang Mai's position as a remote work destination globally and its appeal among digital nomads.

Recent Developments and Emerging Trends in Nimman

Economic modernisation, digital entrepreneurship, tourism growth and urban development initiatives have shaped Nimman's transformation. While these trends have reinforced its status as a commercial and social hub, they have also introduced challenges related to sustainability, affordability and socio-economic disparities.

A significant driver of change has been Nimman's rise as a global hub for digital nomads. Since the early 2010s, Chiang Mai has consistently ranked among the world's top destinations for remote work (Lindner, 2023). The district's youthful demographics, including the student population, digital infrastructure and the expanding network of co-working spaces – such as Yellow Coworking and One Workspace – have fostered an environment conducive to innovation. However, the influx of expatriate professionals has led to rising living costs and altered local business models towards the foreign clientele, raising concerns about economic displacement and cultural transformation.

Government-led urban development initiatives have further accelerated Nimman's evolution. The Chiang Mai SMART city pilot project, launched in 2018 with a budget of 350 million baht (equivalent to 11 million USD), introduced infrastructure enhancements, including the underground installation of electrical and communication cables, improved pedestrian zones and expanded surveillance systems (Thai PBS, 2019). These developments align with Chiang Mai's designation as a UNESCO Creative City, reinforcing its identity as a centre for cultural and technological innovation. The proximity of Chiang Mai University has also played a vital role, fostering collaboration between academia and industry while attracting young entrepreneurs.

Despite these advancements, the COVD-19 pandemic exposed vulnerabilities in Nimman's tourism-dependent economy. At its peak, business revenues reportedly declined significantly, resulting in widespread closures (CityNews, 2021). Limited government support exacerbated financial hardships, particularly for micro- and small-sized enterprises (MSEs). The district is seeing a slow recovery following the return of international travel, largely fuelled by East Asian tourists – albeit not its usual loyal clientele.

Chinese tourism has become a defining force in Nimman's economic landscape. The 2012 Chinese film *Lost in Thailand* significantly boosted Chiang Mai's appeal, with over 1.2 million Chinese tourists visiting the city annually by 2019 (Bindloss, 2020; The Diplomat, 2023). Local businesses have adapted to meet this demand, although evolving travel behaviours – such as a growing reliance on digital recommendations over traditional tour agencies – introduce uncertainties in market trends (Tchou, 2021).

Commercial expansion has further reshaped Nimman's urban fabric. The opening of the Maya Lifestyle Shopping Center in 2014 marked a shift towards upscale retail and entertainment (Sukhsvasti, 2014). Increased investment in boutique hotels, high-end dining and luxury accommodations has reinforced the district's reputation as a cosmopolitan destination. Currently, the Nimman area boasts approximately 112 hotels, 78 restaurants and 66 coffee shops, many of which are clustered around the main road, known as 'Coffee Street.' However, rapid commercialisation has raised concerns about gentrification, as rising property values and rental costs challenge the sustainability of locally owned businesses. While Thai regulations mandate majority local ownership of companies, the increasing influence of foreign capital – particularly from China – continues to reshape the district's economic trajectory (Shoowong, 2023).

Another emerging trend is the rise of medical tourism, aligning with Thailand's broader positioning as a global healthcare destination. Nimman's proximity to leading hospitals, such as Bangkok Hospital Chiang Mai and Sriphat Medical Centre, has made it attractive to medical travellers seeking affordable, high-quality treatments. Thailand ranks among the top medical tourism destinations globally, with procedures costing 40–60 percent less than in Western countries (Noree, Hanefeld & Smith, 2015). This sector is expected to expand, further diversifying Nimman's economic base.

Challenges Facing Nimman

Nimman also faces a series of challenges that threaten its long-term sustainability. While its economic growth and increasing international appeal have positioned it as a key urban centre in Chiang Mai, these developments have also generated structural pressures that require careful analysis. Among the most pressing concerns are infrastructure deficiencies, rising living costs and gentrification, environmental degradation, business volatility and regulatory uncertainties. These challenges affect the district's economic trajectory and have broader implications for its social fabric and urban identity.

Despite ongoing urban improvements, infrastructural constraints continue to impede Nimman's sustainable development. The increasing volume of traffic has placed significant pressure on the district's road network, with congestion around key intersections such as Nimman Square becoming a persistent issue. The limited availability of parking further exacerbates mobility challenges, leading to illegal parking and pedestrian safety concerns. Although Nimman is widely regarded as a pedestrian-friendly area, the absence of a reliable public transportation system forces both residents and visitors to rely heavily on private vehicles, contributing to worsening air quality and inefficient urban mobility. The district's rapid commercialisation has also outpaced infrastructural planning, creating a disconnect between development ambitions and functional urban design. Without targeted interventions to enhance transport infrastructure, pedestrian accessibility and sustainable mobility solutions, these issues will continue to undermine the area's long-term viability.

The escalation of property prices and living costs presents another fundamental challenge, as continued investment in high-end commercial developments and luxury accommodations has contributed to gentrification. Rising rental rates and real estate prices have increasingly marginalised local residents and small business-owners as the district's commercial landscape becomes dominated by enterprises catering to affluent expatriates and international tourists. Many long-established businesses have faced difficulties competing with larger, well-capitalised establishments, resulting in the gradual erosion of the district's original economic and social structure. This trend has reinforced concerns regarding affordability and accessibility as Nimman transitions from a culturally diverse, locally driven district into an area shaped by external investment and global consumer trends. The risk of economic exclusivity underscores the need for policy interventions that preserve the district's affordability while fostering inclusive economic development.

The wider environmental concerns of Chiang Mai also complicate Nimman's development trajectory, as air pollution emerges as one of the most pressing issues confronting numerous Thai regions. The annual smog crisis, primarily driven by agricultural burning in northern Thailand and neighbouring countries, significantly impacts public health and economic activity. Tourism-dependent businesses have been particularly affected, with some hotels experiencing occupancy declines of up to 90 percent during peak pollution months in February and March. While regulatory measures and air quality monitoring systems have been introduced, long-term mitigation requires coordinated regional policies and sustainable agricultural reforms. Beyond air pollution, climate change presents additional risks, as rising temperatures, irregular rainfall patterns and extreme weather events threaten urban infrastructure and resource availability. Adapting to these environmental challenges necessitates resilient urban planning and proactive environmental management strategies.

The economic volatility of Nimman's business environment further highlights the instability of its existing development model. The district's reputation as a trend-driven area has led to fad-chasing businesses, particularly in sectors such as hospitality, retail and co-working spaces. While many entrepreneurs attempt to capitalise on emerging consumer trends, market saturation and shifting preferences have resulted in frequent business closures. This pattern reflects the broader uncertainty surrounding Nimman's economic future, as businesses must constantly adapt to evolving market dynamics while contending with rising operational costs. The challenge lies in fostering a more stable commercial environment that supports long-term investment and business resilience rather than short-term market responsiveness.

Regulatory uncertainties further complicate the district's development trajectory; the increasing influence of foreign investment, particularly from China, has reshaped the district's commercial landscape. While Thai law stipulates that at least 51 percent of any business must be locally owned, the growing presence of foreign capital raises broader questions regarding economic control and long-term sustainability. Balancing investment inflows with local economic agencies will be vital to ensuring that Nimman's development remains in line with broader national and regional priorities.

Stakeholders of Nimman

The ongoing transformation of Nimman into a commercial and tourism centre is driven by a diverse range of stakeholders whose interests significantly affect the district's economic and social environment. Government authorities, private enterprises, investors, residents, digital nomads and tourists all contribute to its development, cultivating a dynamic yet complex environment in which conflicting priorities influence urban policies, commercial strategies and social changes. Comprehending these interactions is crucial for evaluating the district's long-term preferable future and the structural challenges that accompany its swift transformation.

Government institutions play a central role in regulating and facilitating Nimman's expansion, particularly through urban planning and infrastructure development. The Chiang Mai Municipality and the Provincial Government oversee zoning regulations, transportation networks and environmental policies, with initiatives such as the SMART City project aimed at modernising key aspects of the district's infrastructure. Despite such efforts, ongoing issues related to traffic congestion, air pollution and the lack of an integrated public transportation system persist, underscoring the challenges of balancing economic growth with sustainable urban planning. Chiang Mai University further contributes to the district's development as an academic and research institution, fostering a steady population of students and professionals who sustain local businesses. However, the university's increasing exposure to tourism has raised concerns about the encroachment of commercial interests into academic spaces and the preservation of its institutional integrity.

The private sector remains the primary driver of Nimman's economic dynamism, encompassing large-scale commercial developments, multinational corporations and micro-, small- and medium-sized enterprises. Numerous businesses in the hospitality and lifestyle sectors, including boutique hotels, specialty cafés and co-working spaces, have adapted to changing consumer trends. Nonetheless, the volatility of these industries – exacerbated by changing tourist arrivals – brings about economic uncertainty. The legalisation of cannabis has further diversified the district's business landscape, with a growing number of dispensaries emerging in response to regulatory changes, reflecting the broader adaptability of Nimman's commercial sector.

Real estate development has been a defining force in reshaping the district's built environment, with both local and foreign investors playing a significant role in land use transformation. Historically, the Nimmanhaemin family has been a key landholder, leasing properties for commercial expansion. More recently, increasing foreign investment, particularly from China, Japan and South Korea, has driven the proliferation of high-end residential properties, luxury accommodations and upscale retail spaces. While these investments have contributed to economic growth, they have also heightened concerns regarding affordability and the displacement of long-term residents and small businesses. The district's rising cost of living has led to demographic shifts, with local populations increasingly relocating to other areas while expatriates and short-term visitors become the dominant inhabitants.

The social fabric of Nimman reflects the interaction between local residents, expatriates and transient populations such as tourists and digital nomads. Long-standing Thai residents and business-owners remain integral to the district's economic and cultural character, yet the commercial transformation of Nimman has limited affordable housing options, forcing many to commute from surrounding areas. Expatriates and digital nomads have introduced new consumer preferences that have influenced business offerings, while tourists – particularly from China, South Korea and other parts of East Asia – continue to shape market demand. However, the unpredictability of international tourism presents an ongoing challenge, as external factors such as economic downturns and policy changes in source countries can directly impact local businesses.

As Nimman continues to develop, the interactions among these stakeholder groups will shape its future trajectory. While commercial expansion and international investment have driven economic prosperity, unresolved challenges concerning affordability, infrastructure and environmental sustainability remain critical considerations. The district's long-term success will depend on a balanced approach that integrates economic opportunity with cultural preservation and urban resilience. Achieving this equilibrium will require coordinated policy interventions, stakeholder collaboration and sustainable development strategies that ensure Nimman remains both a thriving commercial hub and an inclusive urban space.

Stakeholders' Aspirations and the Preferable Future for Nimman

The aspirations of tourism stakeholders play a vital role in reaching the preferable future. These aspirations establish the connection between the preferable future and the interventions undertaken while considering the contextual issues for destinations and stakeholder groups. Stakeholder aspirations in Nimman represent a complex interplay between social inclusiveness, the pressures of modernisation and market-driven demands. Findings from stakeholder interviews indicate a widespread consensus on the significance of preserving Nimman as the modern face of Chiang Mai while acknowledging the necessity of supporting long-established local businesses in the district.

Stakeholders in the hospitality sector envision an increase in boutique-style hotels, which would enhance the diversity of accommodations available in the Nimman area. A notable proponent of this trend is Mr. Tan Passakornnatee, a prominent entrepreneur and owner of Maya Mall and NuNimman Hotel. His recent investments include the establishment of Oishi Green Tea and several well-known hotels within Nimman. A hotel general manager in the area observed that:

> In the past few years, he (Mr. Tan Passakornnatee) has developed the Nimman area and owns numerous attractions within the district. The stylish and unique boutique hotels are particularly appealing for social media showcases. Consequently, most hotels built in the last five years are smaller boutiques with distinctive rooms. He has played a significant role in this trend.
>
> (Interview 39, April 2024)

These boutique hotels also have the potential to cater to long-term tourists and digital nomads by incorporating diverse and well-equipped working spaces. This need has been echoed by industry professionals, including a front office manager of a mid-sized hotel in the area:

> Several guests asked about options to work in the hotel, but we don't have many options. It would be good to have a hotel and good working space combined.
>
> (Informal conversation with a hotel front office manager, March 2023)

By integrating co-working facilities within boutique hotels, the hospitality sector in Nimman could better accommodate the evolving preferences of modern travellers, particularly remote workers and digital nomads.

Nimman's transformation mirrors global trends in urbanisation, commercial growth and cultural diversity. However, as the district continues to evolve, its future trajectory must be carefully managed to ensure that economic growth does not come at the expense of social inclusivity, cultural preservation and sustainable urban development. Reflecting on an interview, a souvenir shop employee stated:

> The area is turning expensive, and there are now more hotels and shops than before. First, it was only in Nimmanhaemin; now, it's further out. I only come to work in Nimman.
>
> (Interview 36, April 2024)

A preferable future for Nimman envisions a dynamic, multicultural and economically resilient district that balances global connectivity with local identity, mitigates the risks of over-commercialisation and enhances its urban infrastructure to accommodate long-term growth while maintaining a high quality of life for its residents, workers and visitors. Some residents emphasised this demand by stating:

> We would like them (Investors) to add some new places to visit and spend time, interesting venues, not just weed shops, but something appealing for children and locals. We need locations for activities, not solely for eating and drinking, as that won't encourage locals to linger after work.
>
> (Informal conversations with a group of local residents, April 2024)

A defining characteristic of Nimman's preferable future is its emergence as a cosmopolitan urban district that accommodates a diverse range of cultural influences while retaining its distinct urban identity. The 'Salad Bowl' model of multiculturalism provides a conceptual framework for understanding how the district can integrate foreign visitors, investors and expatriates without eroding local traditions and social structures (Berray, 2019). That approach envisions a setting where multiple cultural identities coexist and interact while maintaining their distinctiveness. In this context, Nimman's increasing integration into the global economy, particularly through the influx of East Asian tourists, digital nomads and foreign entrepreneurs, does not necessitate the marginalisation of local businesses or residents. Instead, the district could leverage its international appeal to foster economic diversification, cross-cultural engagement and a sustainable urban economy.

The expansion of tourism, particularly from China, Japan and South Korea, is expected to remain a major driver of economic growth in Nimman. This research indicates that tourism patterns are increasingly shaped by social media trends, with East Asian visitors demonstrating strong preferences for highly aesthetic, experiential and brand-oriented consumer environments. The prevalence of Chinese-language signage, Japanese and Korean restaurants and international retail brands already reflects this shift, and it is anticipated that businesses will continue to align their services with these consumer preferences. However, the long-term sustainability of this model requires strategic diversification to mitigate the risks associated with over-reliance on specific tourism markets. While East Asian visitors currently represent a dominant segment, fluctuations in economic conditions, shifts in geopolitical relations or changes in travel behaviours could introduce vulnerabilities to the district's economic structure. Therefore, a preferable future for Nimman would involve targeted efforts to attract a broader range of international visitors while maintaining a strong appeal for domestic tourists.

Parallel to its development as a tourism hub, Nimman is also expected to consolidate its status as a centre for digital nomads, remote workers and expatriate entrepreneurs. Chiang Mai has already positioned itself as one of the world's leading destinations for location-independent professionals, with Nimman serving as its focal point due to its walkability, high-speed internet infrastructure and vibrant co-working spaces. However, not all digital nomads are content with the current developments. An experienced digital nomad from France stated:

> I don't like the Maya area whatsoever and avoid it at all costs. I prefer the old-school rustic vibe of Chiang Mai's old city side streets. That entire area feels far too bourgeois and congested.
>
> (Interview 6, March 2023)

In other interviews, digital nomads also indicated that they perceived the district as the most dynamic and commercially sophisticated part of Chiang Mai, with an ecosystem that facilitates productivity, networking and lifestyle-oriented consumption. A Dutch digital nomad observed that:

> In the future, I would like to see how traditional styles and advanced technology will be combined together.
>
> (Interview with a digital nomad from the Netherlands, April 2024)

Digital nomads are highly interconnected, facilitating the rapid exchange of information within their community. One of the digital nomads stated that:

> Nimman was recommended to us because it has plenty of lovely cafés to work in, and the overall atmosphere is pleasant and inviting. Our friends had already lived here as digital nomads, and they recommended that we move here as well.
>
> (Interview with a digital nomad from Australia, March 2023)

A preferable future for the district would clearly position Nimman as a hub for long-term remote workers, equipped with modern facilities while preserving its authentic charm and cultural value.

The real estate sector will play a crucial role in shaping the district's socio-economic composition in the coming years. Rising property values, increasing foreign investment and the proliferation of short-term rental accommodations have already contributed to the gradual displacement of local residents, with many relocating to more affordable areas while commuting to Nimman for work. This trend is likely to persist in the foreseeable future, with luxury condominiums, high-end serviced apartments and commercial developments catering primarily to high-income expatriates and investors. While this trajectory aligns with Nimman's evolving status as a globalised commercial hub, it also raises critical concerns about affordability, demographic shifts and the long-term social fabric of the district. Another respondent, a bed-and-breakfast owner, mentioned:

> It is mostly a business area and it's too expensive for local people to live.
>
> (Interview 31, April 2024)

A preferable future would involve a more balanced approach to real estate development, with policies that incentivise the inclusion of affordable housing options, support long-term rental agreements for local residents and regulate the growth of short-term accommodation platforms such as Airbnb. Without such interventions, the risk of gentrification and social stratification may intensify, leading to the erosion of the district's historic character and local community dynamics.

The physical and infrastructural development of Nimman will need to align with its expanding commercial profile. The increasing presence of multinational chains such as Starbucks, H&M and KFC, observed in field research, suggests a consumer preference for globally recognised brands, which in turn influences investment trends and business strategies. While these developments contribute to economic growth, they also accelerate the homogenisation of Nimman's commercial landscape, reducing opportunities for locally owned-businesses to compete. A hotel employee said:

> In the ideal situation, I want to return to the previous composition, in which people can still live here.
>
> (Informal conversation with a hotel receptionist, March 2023)

Another hotel employee also observed that:

> The area changed in terms of people and business increased their prices.
>
> (Informal conversation, March 2023)

In a preferable future, urban planning policies would seek to preserve a balance between international commercial appeal and local entrepreneurship, ensuring that independent businesses retain a foothold in the district's evolving economy. Additionally, the expansion of high-end malls, upscale restaurants and entertainment venues will necessitate infrastructure enhancements to accommodate increasing visitor numbers and commercial density. Addressing existing infrastructural challenges is essential for ensuring the sustainable long-term growth of Nimman. Issues such as traffic congestion, limited parking and the lack of an integrated public transportation system already hinder mobility within the district. As visitor numbers and commercial activity continue to rise, these challenges are expected to worsen. Many respondents highlighted these concerns. A local business-owner emphasised the need for pedestrian-friendly infrastructure:

> Parking is a problem in the area. There is a need for sidewalks and walking paths for people because they are now walking on the street, which is dangerous.
>
> (Interview 24, April 2024)

Similarly, a hotel employee expressed concerns about the limited public transport options:

> I hope it will become a more convenient place in terms of traffic. There are not a lot of buses now.
>
> (Interview 27, April 2024)

A preferable future for Nimman would require coordinated interventions by municipal authorities to improve urban mobility, including expanded pedestrian pathways, enhanced public transportation networks and sustainable transport solutions that reduce the district's dependency on private vehicles. Without such measures, the continued expansion of commercial activity may lead to inefficiencies that undermine the district's accessibility and overall attractiveness.

Ultimately, the preferable future for Nimman is one that balances economic globalisation with cultural and social sustainability. The district will continue to function as Chiang Mai's most cosmopolitan and commercially dynamic neighbourhood, attracting affluent expatriates, East Asian tourists, digital nomads and foreign investors. However, this future will not be without trade-offs. Gentrification, rising living costs and increased corporate influence may alter the district's accessibility for lower-income residents and small business owners, raising concerns about economic inclusivity and local identity preservation.

To mitigate these risks, a strategic approach is required, integrating economic diversification, policy reforms and infrastructure investments. Urban planning initiatives should focus on preserving affordable housing, supporting small businesses and enhancing public infrastructure. Additionally, cultural programming and community engagement efforts should be prioritised to maintain the district's historical and social fabric amidst ongoing transformation. If effectively managed, Nimman's future can serve as a model for sustainable urban development in Thailand, demonstrating how a rapidly globalising district can maintain economic vibrancy while fostering social cohesion and environmental responsibility.

Strategies and Applicable Interventions for a Competitive Urban District

Remaining competitive as a leading commercial and cultural hub in Chiang Mai necessitates dynamic developments in the Nimman area. A preferable future for Nimman requires a carefully coordinated approach that balances commercial expansion with infrastructure development, economic diversification and community engagement. Effective interventions must be designed to address the district's infrastructural deficiencies, mitigate the effects of gentrification, enhance the diversity of business offerings and integrate digital innovation to maintain Nimman's position as an Asian urban district.

One of the most urgent priorities for Nimman's sustainable future is the enhancement of its urban infrastructure, particularly in the areas of mobility and public space management. The rapid increase in visitors, combined with the district's limited road capacity, has led to severe congestion, particularly at high-traffic intersections such as the Nimman–Maya junction. The lack of structured parking facilities and the reliance on private vehicles have further exacerbated these challenges. A multi-level parking facility near major commercial areas would alleviate existing pressures, while the expansion of micro-mobility solutions, including e-scooter and bicycle-sharing programmes, would provide alternative transportation options to reduce vehicle dependency. The construction of pedestrian bridges and the widening of sidewalks in high-density areas would improve walkability and create a more pedestrian-friendly environment. These measures, if implemented in collaboration with municipal authorities and private investors, would strengthen the district's accessibility while aligning with Asian best practices in sustainable urban mobility.

The commercial success of Nimman has been driven largely by tourism and expatriate spending, yet the increasing dominance of international retail chains, cafés and cannabis dispensaries has led to concerns about business homogeneity and market oversaturation. Established long-term business-owners share concerns about this issue, as one restaurant-owner stated:

> I hope for less cannabis stores. It affects the rental fee and I think it's a bit negative, especially with children around.
>
> (Interview with a local Thai restaurant owner, April 2024)

International tourists consider this issue as well. An Asian traveller observed that:

> Include some new spots to visit and enjoy interesting places, not just cannabis shops.
>
> (Informal conversation with a tourist from South Korea, April 2023)

To sustain long-term economic growth, a broader range of business activities must be encouraged. The development of multi-functional cultural spaces would introduce new forms of social and artistic engagement, creating alternative attractions beyond traditional shopping and dining. Underutilised commercial spaces, such as the upper floors of One Nimman and other vacant retail properties, could be repurposed to accommodate art exhibitions, live performances and creative workshops, enhancing the cultural vibrancy of the district. Additionally, the expansion of co-working spaces tailored to digital nomads and students would reinforce Nimman's position as a hub for remote professionals. A postgraduate student remarked that:

> More working spaces would be nice because cafes are quite full and it is also good for CMU students to have a lot of working spaces in the city.
>
> (Informal conversation with a CMU student, March 2023)

The integration of rooftop gardens and small urban parks would contribute to a more sustainable urban landscape, improving the district's overall liveability. These interventions would ensure that Nimman remains not only a commercial centre but also a socially and culturally engaging district.

The rising cost of living and the increasing presence of multinational corporations have led to the gradual displacement of local businesses and long-term residents, reinforcing patterns of gentrification that threaten the district's socio-economic diversity. To counteract these trends, the establishment of a Local Business and Resident Alliance (LBRA) would provide a platform for advocacy and support, bringing together entrepreneurs, property owners and community representatives to safeguard local economic interests. A business incubation programme, supported by municipal authorities and Chiang Mai University, could offer subsidised rental spaces and financial incentives for MSEs, ensuring that locally owned businesses can compete alongside larger corporations. Additionally, initiatives such as street markets, artisan fairs and cultural festivals would provide economic opportunities for small vendors while preserving the district's cultural heritage. Current residents seek greater diversity in local businesses. One resident remarked:

> There needs to be places to do something such as activities and not only eating and drinking.
>
> (Informal conversations with residents, March 2023)

Addressing the housing affordability crisis is equally critical. Encouraging real estate developers to include affordable housing units in new projects, potentially through regulatory incentives or tax benefits, would help mitigate the displacement of long-term residents. These measures would foster a more inclusive urban environment, ensuring that economic development does not come at the cost of social cohesion.

As Nimman continues to attract digital nomads and social-media-driven tourists, leveraging digital integration will be essential to sustain its competitive advantage. Establishing a digital nomad-friendly business certification programme would recognise establishments that cater specifically to remote workers, offering reliable internet access, ergonomic workspaces and networking opportunities. Certification from the Tourism Authority of Thailand would enhance the visibility of participating businesses, positioning them as preferred locations for digital professionals. Additionally, the growing influence of social media on tourism behaviour highlights the need for businesses to integrate visually compelling design elements into their branding and architecture. Encouraging establishments to incorporate distinctive murals, rooftop terraces and aesthetically engaging storefronts would enhance Nimman's appeal as a photogenic, trendsetting destination. A digital marketing grant programme, administered through the LBRA in collaboration with private investors, would further support local entrepreneurs in expanding their online reach, allowing them to remain competitive in an increasingly digitalised economy.

These interventions would provide a framework for realising a preferable future in which Nimman remains economically vibrant, socially inclusive and environmentally sustainable. Infrastructure improvements would enhance mobility and accessibility, commercial diversification would expand the range of economic and cultural activities available, community-led initiatives would protect local businesses and residents from displacement and digital innovation would ensure that Nimman retains its global relevance. These measures, if effectively implemented through coordinated efforts between municipal authorities, business leaders and community stakeholders, would ensure that Nimman's future remains aligned with broader principles of sustainable urban growth.

Conclusion

This study of the preferable future of Nimman has highlighted the district's rapid transformation into an internationally oriented commercial and cultural hub, shaped by the increasing influence of East Asian tourists, digital nomads and foreign investors. While this growth has bolstered its economic importance, it has also brought about structural challenges that jeopardise its long-term sustainability. The district faces critical issues related to infrastructure strain, excessive commercialisation, gentrification and the demands of a digital economy that favours high-visibility commercial spaces over locally entrenched businesses. Without strategic interventions, Nimman risks becoming an exclusive, transient and commercially saturated area that primarily caters to short-term visitors and wealthy expatriates, thereby marginalising its long-term residents and local entrepreneurs. The findings of this study demonstrate that addressing these challenges require a coordinated approach that enhances urban infrastructure, promotes commercial diversification, strengthens local community engagement and integrates digital strategies that sustain economic competitiveness while preserving social inclusivity.

One of the most pressing challenges identified in this research is the district's worsening infrastructural inefficiencies, particularly in relation to traffic congestion, mobility constraints and inadequate pedestrian accessibility. The rapid expansion of commercial enterprises and rising visitor numbers have placed unsustainable pressure on Nimman's road network, leading to daily bottlenecks that reduce both the liveability and functionality of the district. The proposed interventions respond to this challenge through the introduction of a multi-level parking facility, micro-mobility solutions and the redesign of the Nimman–Maya junction. These measures would improve urban mobility, reduce traffic congestion and enhance the pedestrian experience, ensuring that Nimman remains an accessible and efficient commercial district. Without such interventions, continued growth will further strain existing infrastructure, limiting the district's long-term viability.

The second major challenge is the increasing dominance of multinational businesses and speculative real estate investments, which have driven up property values and displaced local businesses. The swift commercialisation of Nimman has resulted in an increasingly homogenised economic landscape, where independent Thai-owned businesses find it challenging to compete with well-capitalised foreign brands. The proposed LBRA would provide a direct response to this issue, serving as a platform for local entrepreneurs to advocate for their economic interests. By implementing subsidised rental programmes for small businesses, supporting local markets and promoting cultural festivals, this initiative could enable locally embedded enterprises to remain viable alongside larger corporate investments. Additionally, regulatory frameworks that control the proliferation of short-term rental accommodations would help stabilise housing costs, preventing the large-scale displacement of long-term residents. These measures would address the risks associated with gentrification, and help ensuring that Nimman's transformation does not come at the cost of social and economic exclusivity.

Another key challenge identified in this study is the increasing dependence on social-media-driven tourism, which has led to a highly aestheticised and consumption-oriented urban experience. While digital visibility has fuelled Nimman's economic expansion, it has also contributed to commercial saturation, where businesses prioritise marketability over long-term sustainability. The proposed interventions in digital engagement respond to this challenge by providing small businesses access to digital marketing grants and encouraging the integration of distinctive urban design elements that align with social-media-driven consumer behaviour. These measures would allow independent enterprises to remain competitive in an evolving digital marketplace while ensuring that commercial development did not erode the district's urban identity.

The findings of this research underscore the need for a balanced and strategic approach to urban planning, where commercial growth is managed in a way that supports economic resilience, social inclusivity and environmental sustainability. Nimman's future success will depend on how effectively municipal authorities, private investors and community organisations implement such proposed interventions. If managed well, the district has the potential to serve as a model for sustainable urban development in Thailand, demonstrating how global economic integration can coexist with local cultural and economic preservation. By addressing infrastructural inefficiencies, mitigating gentrification, diversifying commercial offerings and adapting to digital economic trends, Nimman can continue to thrive as a dynamic, future-oriented urban hub while ensuring that its development remains inclusive and sustainable.

While this study has provided a comprehensive analysis of its current landscape and preferable future, several areas require deeper investigation to ensure a sustainable and inclusive development trajectory. Future research should focus on the long-term socio-economic impacts of globalisation on Nimman, the effectiveness of urban planning interventions and the role of digital economies in shaping the district's commercial environment.

One critical area for future study is the socio-economic impact of increasing foreign investment and real estate speculation on local businesses and residents. While this study has identified gentrification and rising property values as significant challenges, further research is needed to assess the extent to which these factors contribute to the displacement of small businesses and long-term residents. Longitudinal studies tracking changes in land use, business ownership patterns and demographic shifts would provide valuable insights into how urban policies can mitigate the negative effects of rapid commercialisation while preserving local economic agency.

Another important research avenue is the evaluation of infrastructural interventions aimed at improving mobility and urban accessibility in Nimman. The proposed measures, including multi-level parking solutions, micro-mobility services and traffic system redesigns, require an empirical assessment to determine their feasibility and long-term impact. Future studies should include cost–benefit analyses, traffic simulations and pedestrian movement studies to assess the effectiveness of these strategies in reducing congestion and enhancing walkability. Comparative studies examining best practices in sustainable urban mobility from similar commercial districts in Southeast Asia could offer additional policy recommendations tailored to Nimman's unique urban context.

Further research is also needed on the long-term sustainability of Nimman's digital economy and social-media-driven tourism. While digital visibility has played a central role in shaping the district's commercial success, questions remain about the shifting preferences of East Asian tourists that would help businesses and policymakers develop adaptive strategies that ensure long-term competitiveness.

Finally, given the increasing pressures of urbanisation on Nimman's infrastructure and resources, an interdisciplinary approach incorporating environmental sustainability into future research would be crucial. Studies exploring sustainable architectural design, green urban planning and the ecological impact of commercial expansion could provide essential guidelines for integrating environmental responsibility into the district's development. Research on waste management, energy consumption and carbon emissions associated with increased tourism and commercial activity would also contribute to a more holistic understanding of how Nimman can achieve sustainable growth without exacerbating environmental degradation.

References

Berray, M. (2019). A critical literary review of the melting pot and salad bowl assimilation and integration theories. *Journal of Ethnic and Cultural Studies*, *6*(1), 142–151.

Bindloss, J. (2020). *Lost in Thailand*: The Chinese take on Chiang Mai. Culture Trip. Available at https://theculturetrip.com/asia/thailand/articles/lost-in-thailand-the-chinese-take-on-chiang-mai (Accessed 26 February 2025).

CityNews. (2021). Businesses are giving up. Available at www.chiangmaicitylife.com/citynews/local/businesses-are-giving-up/ (Accessed 26 February 2025).

Flatio. (2024). Top 10 cities for digital nomads and expats in 2024. Available at www.flatio.com/blog/top-10-cities-for-digital-nomads-and-expats-in-2024 (Accessed 26 February 2025).

Lindner, J. L. (2023). Must-know digital Nomads statistics. GITNUX. Available at https://gitnux.org/digital-nomads-statistics/ (Accessed 26 February 2025).

Mostafanezhad, M., & Promburom, T. (2018). '*Lost in Thailand*': The popular geopolitics of film-induced tourism in northern Thailand. *Social & Cultural Geography*, *19*(1), 81–101.

Noree, T., Hanefeld, J., & Smith, R. (2015). Medical tourism in Thailand: A cross-sectional study. *Bulletin of the World Health Organization*, *94*(1), 30.

Shoowong, M. (2023). Chiang Mai frets over foreign nominees. *Bangkok Post*. Available at www.bangkokpost.com/business/2512036/chiang-mai-frets-over-foreign-nominees (Accessed 26 February 2025).

Sukhsvasti, U. (2014). It's glam up north. Available at www.bangkokpost.com.lifestyle/394023/it-glam-up-north (Accessed 26 February 2025).

Tchou, L. (2021). The Changing travel habits of Chinese tourists. CUHK Business School. Available at https://cbk.bschool.cuhk.edu.hk/the-changing-travel-habits-of-chinese-tourists/ (Accessed 26 February 2025).

Thai PBS. (2019). Nimmanhaemin Rd. chosen as pilot to transform Chiang Mai into a SMART city. Available at www.thaipbsworld.com/nimmanhaemin-rd-chosen-as-pilot-to-transform-chiang-mai-into-a-SMART-city/ (Accessed 26 February 2025).

The Diplomat. (2023). Asia travel hotspots quiet as Chinese tourists stay away. *The Diplomat*. Available at https://thediplomat.com/2023/01/asia-travel-hotspots-quiet-as-chinese-tourists-stay-away/ (Accessed 26 February 2025).

10 The Preferable Future of the Community-Based Tourism Village, Hmong Doi Pui, Chiang Mai, Thailand

Introduction

Chiang Mai area has emerged as a prime location for community-based tourism (CBT) in Thailand, attracting travellers interested in authentic and sustainable tourism experiences. The area boasts a rich cultural heritage, varied natural scenery and significant Indigenous and ethnic minority groups, creating a perfect environment for CBT projects. These projects offer engaging activities, including homestays, cultural immersion programmes and eco-tourism opportunities, addressing the increasing desire for responsible tourism travel.

Boonratana (2010) defines CBT as a socially, economically, environmentally and culturally responsible form of tourism. It allows visitors to engage with local and Indigenous communities while appreciating their cultural and natural heritage. A key feature of CBT is that tourism resources, products and services are developed and managed with the active participation of the host community, ensuring that benefits – whether tangible or intangible – are fairly distributed among its members (Scheyvens, 2002). According to the United Nations Development Programme (UNDP, 2022), CBT has gained recognition as an effective strategy for strengthening local economies, conserving cultural and natural heritage and supporting marginalised communities in Thailand.

Despite its potential, CBT has also faced considerable criticism for its practical limitations across various Asian destinations (Han, Wu, Huang & Yang, 2014; Zhu, Zhan & Li, 2021). In Thailand, its implementation is frequently impeded by inadequate community capacity, deeply entrenched structural inequalities in decision-making and tensions between centralised and decentralised governance (Ditta-Apichai, Sroypetch & Caldicott, 2024). These challenges lead to power asymmetries, where local communities encounter limited access to resources, gaps in tourism knowledge and skills and exclusion from essential decision-making processes. Consequently, in worse-case scenarios, rather than fostering empowerment, CBT can exacerbate existing disparities, restricting communities' capacity to fully engage in and benefit from tourism development.

This chapter envisions a preferable future for Hmong Doi Pui village, ensuring that tourism development is in harmony with sustainability principles while

DOI: 10.4324/9781003570240-13

improving the livelihoods of its inhabitants. To achieve this, the study investigates the effectiveness of CBT activities in the village and their potential to promote economic resilience, cultural preservation and sustainable environmental stewardship.

Several successful CBT initiatives in the Chiang Mai area demonstrate the potential of CBT in fostering sustainability development. Among them, Mae Kampong village stands out as a model of CBT. Located an hour from Chiang Mai, Mae Kampong offers visitors the chance to experience daily village life, from staying with local families to learning about coffee production, traditional herbal medicine and local handicrafts. Another notable CBT destination is Doi Inthanon National Park, home to Thailand's highest peak. The park provides eco-lodges and homestays, allowing visitors to experience hill tribe culture, explore nature trails and engage in conservation initiatives. These two destinations illustrate how CBT can empower local communities while maintaining cultural and environmental integrity.

Hmong Doi Pui has been practising CBT for the past two decades. Historically, the village has undergone social and economic transformations, shifting from opium cultivation to sustainable agriculture and later adopting tourism as a key economic driver (Betgem, 2022). However, the village's reliance on tourism also presents challenges, particularly in maintaining cultural authenticity, environmental sustainability and economic resilience. A key challenge currently facing the village is the disconnection between visitor expectations and the actual tourism experience. Many visitors arrive seeking authentic cultural interactions; however, their experiences are often restricted to shopping for souvenirs and brief, superficial encounters with the community. The short duration of visits, often constrained by structured tour schedules, further prevents deeper engagement with local traditions and ways of life. Additionally, language barriers, a lack of pre-arranged interactive activities, and the dominance of commercial tourism models have contributed to a fragmented visitor experience, where the cultural richness of the village is not fully communicated or appreciated.

This study focuses on identifying a vision for Hmong Doi Pui's future by proposing strategic interventions. By leveraging its rich cultural heritage, natural surroundings and traditional knowledge, the village has the potential to develop a sustainable and culturally immersive tourism experience. Through a forward-looking approach, this study seeks to ensure that tourism in Hmong Doi Pui contributes positively to both the local community and the broader sustainable tourism landscape in Thailand.

Historical Context and Tourism Development in Hmong Doi Pui

Hmong Doi Pui village, situated at 1,300 metres above sea level in the Doi Suthep mountains, has experienced significant social and economic transformations, primarily driven by the growth of tourism as a key economic force. The village's transition into a tourist destination began in 1960, following a visit from King Rama IX, which led to the establishment of the first visitor-oriented shop and marked the formal introduction of tourism-driven economic activities. Throughout the 1970s and early 2000s, the village saw a swift rise in visitor numbers due to international tourists' interest in Chiang Mai, peaking in 2008, when 5,000 visitors arrived each day during the high tourist season. While tourism has provided economic opportunities and improved local livelihoods, its rapid expansion has also brought about environmental and social challenges, including waste pollution, fire hazards and concerns regarding cultural commodification. More recently, the COVID-19 pandemic severely disrupted the village's tourism industry, resulting in temporary closures, income losses and economic uncertainty.

Despite these fluctuations, tourism remains the backbone of the village's economy, with most businesses focusing on traditional crafts, clothing, herbal medicines and cultural education for visitors. To safeguard cultural authenticity and community ownership, the village has implemented regulations restricting shop ownership to Hmong residents and continues to promote policies aimed at preventing cultural exploitation. Efforts are also being made to encourage younger generations who have pursued education in urban areas to return and contribute to the village's sustainable development (Chiang Mai Guideline, 2018).

Beyond tourism, Hmong Doi Pui possesses a rich socio-economic history intricately linked to changes in agricultural practices. Founded over 80 years ago with just seven homes, the village initially depended on opium cultivation, which was prevalent in Northern Thailand during the Vietnam War. This practice, however, led to severe economic and social issues, such as poverty, high child mortality and low life expectancy. A significant change took place in 1963 when King Rama IX conferred Thai nationality upon the Hmong people and promoted lychee farming as a viable alternative. By 1972, opium farming had been entirely eradicated, paving the way for sustainable agricultural practices and a variety of economic activities. Over time, improvements in infrastructure and favourable government policies enabled certain Hmong communities to relocate to lower altitudes, enhancing their access to education, healthcare and economic opportunities.

Despite these changes, Hmong Doi Pui has maintained its unique cultural identity, marked by a communal and clan-based social structure. The community in the village, known as the Hmong people, migrated initially from Laos across the Mekong River and has preserved unique cultural practices, including rituals, beliefs in household spirits and traditional craftsmanship. Today, the village consists of approximately 360 houses, with an estimated population of 2,000–2,500 residents (Betgem, 2022). Educational initiatives have played a crucial role in cultural preservation, particularly through the establishment of the village's first school by King Rama IX. The school continues to evolve, with plans to develop a specialised curriculum aimed at safeguarding the Hmong language and traditions.

Another major transformation in the village's economic landscape has been the shift from opium cultivation to environmental conservation and sustainable agriculture. The surrounding areas, once dominated by opium farms, have been successfully reforested and converted into fruit plantations, reinforcing the community's commitment to sustainable land use. This area is currently utilised as a hiking route known as the *opium trail* and is offered as a day activity visit.

While tourism has proven vital for the village's economy, addressing its associated challenges – such as over-commercialization, environmental degradation and uneven community benefits – remains critical.

The Cultural and Economic Assets of Hmong Doi Pui Village

Hmong Doi Pui village, nestled in the forests of Doi Suthep-Pui National Park, is accessible via a single winding road. The village provides visitors with a glimpse into Hmong culture, traditional craftsmanship and the natural beauty of the surrounding mountains. Despite its expansive landscape, much of the visitor experience is shaped by a structured flow through designated areas, where signs and pathways guide tourists towards specific attractions.

Upon arrival, visitors are first introduced to the village through its central square and parking area, a bustling entry point where the rhythms of daily life intersect with the influx of tourism. From here, many are drawn to the market street, lined with souvenir shops, small cafés and local eateries, where merchants display a familiar selection of handmade textiles, silver jewellery, dried fruits, wine and locally sourced coffee. While many of these products are sourced collectively from the region, each vendor operates independently, competing for customers passing through the main walkways. Amid this commercial activity, a few businesses distinguish themselves, such as an artisan shop selling handcrafted hemp earrings and unique Hmong clothing, offering visitors a chance to purchase items that reflect the individuality of their makers.

Beyond the market, the village unfolds into more tranquil spaces, where visitors encounter a small museum showcasing Hmong history, an adjacent flower garden and a network of scenic pathways leading to further cafés and a waterfall. These attractions, requiring a modest entrance fee, provide visitors with a slower, more contemplative experience, inviting them to pause, take in the mountainous landscape and appreciate the region's floral diversity. While these sites serve as economic assets, drawing in tourists willing to pay for entry, they also reinforce the structured nature of the visitor experience – most tourists remain within these well-defined zones, rarely venturing beyond the established pathways into the deeper, everyday life of the village.

For those interested in a prolonged visit, Hmong Doi Pui features three homestays that can host up to 38 guests in private rooms equipped with modern amenities. Nevertheless, these accommodations are not easily noticeable, as they are located away from the primary tourist attractions. As a result, most visitors depart by early evening, allowing the village to return to its tranquil pace once the day's tourism activities conclude. This trend means that very few visitors experience the community's life after dark, missing the opportunity to observe how locals interact, how families gather and how traditions are maintained beyond the familiar context of tourism.

Culture plays a central role in CBT, serving as both a community asset and a key component of the tourism experience. Boonratana (2010) emphasises that host communities retain their traditional way of life and cultural practices, which are of intrinsic interest to visitors (p. 284). This suggests that tourists expect cultural authenticity throughout their stay, reinforcing the idea that local traditions are not only preserved but also actively showcased. This perspective is particularly relevant to Hmong Doi Pui, where cultural heritage – ranging from traditional crafts and rituals to agricultural practices and social structures – is deeply embedded in the village's identity. Recognising a growing interest in experiential tourism, a small number of villagers have begun offering hands-on cultural workshops, enabling visitors to engage with traditional crafts, textiles and other aspects of Hmong heritage. These initiatives are part of the D-Hope project, a Community-Based Entrepreneurship Promotion Project supported by a national government initiative led by the Japan International Cooperation Agency (JICA) to promote sustainable CBT. By fostering local entrepreneurship and leveraging cultural resources, the project aims to generate social and economic benefits for rural communities. It has been implemented in Chiang Mai, Ranong, Trang, Chanthaburi, Chonburi, Surin, Nakhon Phanom, Mukdahan and Lamphun, reflecting a broader shift towards immersive and locally driven tourism experiences. With a strong focus on homestay programmes and interactive cultural exchanges, D-Hope aims to enhance visitor engagement while bolstering local economies. While still in its early stages, the programme signals a shift towards more interactive and locally driven tourism experiences, with future plans to introduce expanded homestay programmes and immersive cultural exchanges. The emergence of these initiatives reflects a broader motivation within the village – one driven not only by economic opportunities but also by a shared desire to preserve and convey Hmong identity to those who visit.

As CBT villages are increasing around Chiang Mai, Hmong Doi Pui faces rising competition from both new village destinations and more traditional tourist sites. Nevertheless, the village possesses distinct advantages that differentiate it from others. Its locally sourced coffee, handcrafted artisan products and picturesque location within the national park significantly enhance its attractiveness as a destination where culture, nature and local enterprise converge. These attributes not only play a crucial role in drawing visitors but also act as sources of community pride, fostering a collective commitment to highlighting the village's unique heritage. A considerable number of local residents perceive tourism as an opportunity by which to enhance their English language proficiency, engage with a global audience and remain attuned to evolving visitor expectations. The D-Hope project, in particular, exemplifies a progressive willingness to explore innovative forms of engagement, indicating a future wherein tourism in the village becomes increasingly participatory, educational and culturally immersive.

The visitor base of Hmong Doi Pui is varied, encompassing solo travellers, families, tour groups and expatriates. Some arrive with a keen cultural curiosity, eager to observe Hmong traditions, agricultural practices and village life. Others are attracted to the tranquil mountain landscape, seeking a peaceful retreat from Chiang Mai's urban intensity. For many Thai visitors, the village is a haven for enjoying nature and relaxation, while international tourists often show interest in learning about Hmong culture through workshops and direct engagement with locals. The village's culinary allure, especially its reputation for high-quality coffee, also contributes to its appeal, with many identifying it as one of the key highlights of their trip.

For some, the village's remoteness enhances its mystique and allure. Some visitors arrive spontaneously, incorporating the village into their itineraries after noticing roadside signs or recommendations for tour packages. This combination of intentional cultural immersion and incidental discovery highlights the necessity of managing visitor expectations, ensuring that the village's image aligns with the experience it offers. Whether visitors come for a carefully planned cultural exchange or simply stumble upon it as an additional excursion, their overall perception of the village is influenced by how well the experience matches their initial expectations.

Reaching Hmong Doi Pui is relatively straightforward, with various transport options catering to different visitor segments. Chinese tourists tend to book organised tours, either directly through tour operators or indirectly via hotels and accommodations. Thai visitors, in contrast, often arrive in private vehicles, taking advantage of their familiarity with the region and the flexibility of personal transport. Many international travellers, particularly independent tourists, prefer renting scooters or cars, allowing for self-paced exploration and spontaneous detours. Meanwhile, songthaews (red trucks) provide an affordable public transport option, offering shared rides or private hires with guided services for those seeking a deeper understanding of the village's culture.

Despite the variety of transportation options available, most organised tours from Chiang Mai regard Hmong Doi Pui as a secondary stop. They typically include it in itineraries alongside visits to Wat Phra That Doi Suthep and other regional landmarks. This structured framing influences how the village is experienced, with fixed time slots and predetermined points of interest shaping visitor engagement. Consequently, many visitors do not venture beyond the primary tourist sites, reinforcing a somewhat superficial encounter with the village.

Methodology

This study employs a mixed-methods approach, integrating both qualitative and quantitative research tools to examine the dynamics of tourism stakeholders in Hmong Doi Pui village. Data collection was conducted over two consecutive weeks in April 2024 (April 10–24) through an agile ethnographic fieldwork approach, supplemented by netnography, participant observation and stakeholder interviews. Prior to fieldwork, desk research provided historical and contextual insights into the village, while online visitor reviews were analysed to assess tourist perceptions. Photovoice was employed to capture how villagers perceive their roles within the tourism landscape, and a small visitor survey, along with informal interviews, gathered perspectives from local tourism entrepreneurs and visitors. Additionally, public documents, media sources, promotional materials from tour operators, maps and photographs were collected with the support of Thai tourism scholars. The primary dataset consists of 32 recorded interviews and informal conversations with key informants across the public and private sectors, including tour operators, NGO executives and local entrepreneurs.

Data analysis involved a multi-step thematic approach. Observations and interviews were transcribed and categorised into six thematic clusters to identify recurring patterns and insights into both the opportunities and challenges facing CBT in the village. Photovoice analysis provided a visual narrative of local aspirations, while visitor surveys highlighted key concerns such as language barriers, over-commercialisation and limited interaction with locals. Findings were further validated through expert confirmation interviews, ensuring alignment with industry trends and community needs. The final synthesis of these data sources informed the development of feasible interventions aimed at fostering a more authentic and sustainable tourism experience in Hmong Doi Pui village.

Challenges and Opportunities for Destination Management in Hmong Doi Pui Village

The Hmong Doi Pui village faces several challenges in aligning its tourism offerings with its cultural and natural assets, particularly in meeting visitor expectations for authenticity. A notable concern is the issue of commercialisation, which has led to a disconnect between visitors' expectations and their actual experiences. Many tourists anticipate authentic cultural interactions; however, their experiences are often influenced by an abundance of souvenir shops selling identical products and performative cultural expressions, such as donning Hmong attire for staged photographs. This gap in promotional activities might lead to visitor dissatisfaction due to the shallow engagement with Hmong traditions limited.

Another key challenge is the lack of awareness regarding homestay accommodations, which hinders opportunities for extended stays and deeper cultural immersion. The language barrier further complicates meaningful exchanges between visitors and locals, reducing the potential for transformative cultural experiences. Additionally, the limited capacity of homestays, combined with the rushed itineraries imposed by travel agents, restricts visitors' ability to fully explore the village beyond its predefined tourist zones. The prevailing model of brief, structured visits prevents tourists from developing a deeper appreciation of the village's way of life, reinforcing the perception of Hmong Doi Pui as merely a quick stop within a larger itinerary rather than a destination worth lingering in.

However, despite these challenges, Hmong Doi Pui holds significant opportunities to reposition itself as a destination that encourages slower, more immersive experiences. The village's unique selling points (USPs) – coffee, culture and its natural environment – can serve as key drivers for sustainable tourism development. By leveraging these assets, Hmong Doi Pui could shift from a superficial tourism model to a more meaningful and experience-driven approach, aligning with emerging visitor preferences for cultural immersion and hands-on activities.

One of the most promising opportunities lies in the concept of 'slow life' tourism, a philosophy that aligns seamlessly with the natural rhythm of the village and its cultural traditions. 'Slow Life' tourism emphasises mindfulness, presence and a deep connection to the environment and community. Hmong Doi Pui exemplifies this lifestyle through its agriculture, traditional crafts and community living. This existing but underexploited potential could be developed into a compelling tourism experience that caters to the increasing desire for peace, mindfulness and genuine connections engagement.

By embracing 'Slow Life' tourism, visitors can disconnect from the fast-paced urban world and reconnect with a simpler, more intentional way of living. This concept is not merely a marketing tool but a genuine reflection of the Hmong people's approach to life, seen in their hospitality, agricultural rhythms and appreciation for craftsmanship. Whether through meditation, traditional coffee preparation or learning the intricacies of Hmong textile weaving, visitors could be encouraged to fully immerse themselves in the process rather than focusing solely on the end product. This approach would enhance the visitor experience while also providing economic and cultural value to the local community.

These challenges and opportunities highlight the careful balance between maintaining cultural heritage and meeting tourism demands. As the village manages these complexities, it has the potential to enhance its offerings in a manner that optimises both economic gains and cultural sustainability, ensuring that tourism development supports the community's long-term well-being.

The Role of Stakeholders in CBT Coordination at Hmong Doi Pui: A Power Relations Analysis

The development and management of CBT in Hmong Doi Pui involves a diverse range of stakeholders, each possessing varying levels of power, influence and dependency. These stakeholders can be categorised into four key groups: (1) the local community, (2) market-linked organisations (MLOs) such as tour operators and NGOs, (3) governmental institutions and (4) academics and external experts. Understanding the power dynamics between these groups is essential for identifying the challenges and opportunities in the sustainable development of CBT in the village.

Stakeholders in Hmong Doi Pui can be organised along a power hierarchy, where decision-making influence is concentrated at the top among governmental organisations and tour operators, while local communities occupy the lowest level of influence.

At the highest level of power and authority, governmental organisations such as the Designated Areas for Sustainable Tourism Administration (DASTA) establish the overall direction for CBT through policy-making, funding allocation and regulatory oversight. These entities hold definitive stakeholder status, as they control the legal and financial framework within which all other stakeholders must operate. Despite their considerable influence, there is a growing mistrust among local communities due to concerns over top-down decision-making, limited community involvement and the prioritisation of CBT expansion over long-term sustainability.

MLOs, primarily consisting of tour operators and non-governmental organisations (NGOs), sit just below government entities in the power structure. Tour operators wield considerable influence as they determine which CBT villages are included in their itineraries, directly impacting local tourism revenue. They act as gatekeepers to the visitor market, controlling the flow of tourists and the narrative presented about each village. This significant degree of influence creates a dependency, where villages not included in tour packages may struggle to attract visitors and sustain their tourism initiatives. Furthermore, as new tourism markets emerge – such as MICE (meetings, incentives, conferences and exhibitions) tourists, wellness travellers and digital nomads – tour operators will play a key role in shaping product diversification. However, the increasing number of operators entering the CBT market has led to heightened competition, compelling operators to differentiate their offerings while balancing their business interests with community needs.

NGOs occupy a more complex and nuanced role within the stakeholder network. They are seen as dependent stakeholders, possessing a strong urgency to promote sustainable CBT development but lacking direct authority over decision-making. NGOs act as intermediaries between the government, local communities and tour operators, advocating for community interests, capacity building and sustainability initiatives. However, their dependence on external funding creates vulnerabilities, as financial limitations compel them to prioritise certain villages over others, resulting in some communities being unsupported. In the long term, the role of NGOs may evolve as local communities grow more self-sufficient, forming their own organisations to access funding directly. Nevertheless, conflicts between NGOs, government bodies and tour operators persist as each stakeholder group has distinct and sometimes opposing objectives.

At the bottom of the power hierarchy, the local community, including village leaders and residents, holds minimal direct influence in CBT decision-making. Although villagers are vital guardians of Hmong culture and traditions, their involvement in shaping tourism efforts tends to be reactive rather than proactive. Village leaders do wield some power, especially in representing the community and negotiating with outside parties, yet they often lack the necessary legitimacy and urgency to foster independent decision-making. The wider community is seen as demanding stakeholders due to their strong needs and expectations for tourism benefits, but they have limited ability to advocate for structural changes. A significant challenge is that community members depend on external actors – like tour operators, NGOs and government agencies – for tourism facilitation, marketing and visitor engagement, leaving them particularly exposed to external decisions that are out of their control.

Challenges in Multistakeholder Collaboration

Several tensions exist within the stakeholder network, particularly regarding collaboration, power imbalances and conflicting priorities.

Government versus Community Trust Issues

The government's push for CBT expansion often overlooks the long-term viability of individual villages, leading to intense competition among CBT destinations. Many villages struggle to remain financially sustainable, fostering growing scepticism and distrust towards governmental intervention. Without proper community engagement, there is a risk that villages like Hmong Doi Pui may experience tourism fatigue or resource depletion, undermining the original goals of CBT.

NGO Funding and Neutrality Challenges

NGOs serve as bridges between various stakeholders; however, their reliance on external funding can introduce biases into their decision-making. The necessity to secure financial support occasionally results in prioritising donor expectations over authentic community needs, which can undermine the trust between NGOs and villagers.

Tour Operator Dependency and Market Control

Tour operators have a considerable impact on which villages are part of tourism routes, putting CBT communities in a vulnerable position. A village that does not meet the expectations of tour operators may find it difficult to draw visitors, despite its genuine cultural significance. Furthermore, intense competition among tour operators has led to decisions based primarily on pricing, favouring brief, standardised tours instead of deep, culturally respectful experiences.

Limited Community Participation in Decision-Making

While the success of CBT depends on community involvement, the existing power structure places villagers at the bottom of the decision-making process. Limited access to language skills, marketing knowledge and tourism training further limits their ability to assert control over their tourism future. Although initiatives such as homestays and cultural workshops have emerged, their impact remains constrained by external influences and structural limitations.

Opportunities for a More Balanced Power Dynamic

Despite these challenges, opportunities exist to establish a more equitable stakeholder structure within Hmong Doi Pui's CBT development.

Strengthening Community Leadership and Local Entrepreneurship

Empowering local residents through training programmes in tourism management, marketing and language skills could help shift decision-making power back to the village. Encouraging CBT initiatives, such as self-managed homestays and craft workshops, could reduce reliance on external actors and enable the village to directly benefit from tourism revenue.

Reforming Tour Operator Practices

Encouraging tour operators through public organisations to embrace more sustainable and ethical business practices – such as extended stays, small-group travel and deeper cultural immersion – could enhance the quality of visitor experiences while minimising negative impacts on the community.

Fostering More Inclusive Policymaking

Governmental organisations such as DASTA could involve communities more actively in policy discussions, ensuring that local perspectives are incorporated into tourism planning. By enabling direct communication between community leaders and policymakers, a more collaborative governance approach could be established.

Diversifying Funding Sources for NGOs and Communities

Mitigating the overreliance on external funding through the establishment of community-led organisations can improve financial independence and enable NGOs to transition towards capacity-building roles rather than direct funding allocation.

By tackling these power disparities and encouraging increased community involvement, Hmong Doi Pui could shift from a tourism-reliant model to a self-sustaining, culturally respectful approach. This transition would help ensure that CBT practices remain economically viable, socially equitable and environmentally sustainable for the long term.

Preferable Future of Hmong Doi Pui: A Vision for Reclaiming its Role Model Status as a CBT Village in Chiang Mai

The preferable future for tourism in Hmong Doi Pui village envisions a harmonious balance between preserving culture, enhancing community well-being, and offering immersive experiences for visitors. This vision, grounded in the aspirations of local residents and tourists, aims to establish the village as a destination where cultural exchange, stunning nature and the spirit of slow living come together to provide a fulfilling and refreshing experience for everyone who visits.

At the heart of this vision lies the community's profound connection to its cultural heritage and natural surroundings. The villagers take pride in sharing their way of life through coffee, cuisine, handicrafts and traditional knowledge, ensuring that tourism serves as a means of cultural expression rather than commodification. The visitor experience, in turn, is defined by a sense of tranquillity and authenticity, where time slows down and meaningful interactions take precedence over rushed sightseeing. As the head of the village articulates:

> Tourists enjoying the slow life to get to know the culture and make deeper connections.
>
> (Interview 9, April 2024, encapsulates the village's commitment to fostering genuine engagement.)

In this future-oriented model, Hmong Doi Pui would be setting itself apart from other CBT villages by embracing the philosophy of slow living, which is deeply ingrained in Hmong traditions and beliefs. The village offers visitors the opportunity to disconnect from the stresses of modern life and reconnect with nature and people. This ethos is reflected in both daily interactions and structured experiences, such as learning the intricate process of coffee cultivation, engaging in craft-making workshops or simply sharing moments of conversation with locals. The deliberate, unhurried pace of life allows visitors to fully absorb the rhythms of the community, fostering a sense of personal rejuvenation and cultural appreciation.

This vision aligns with the increasing global movement towards mindful and sustainable travel, where visitors seek deeper connections with the places they explore. As noted by a café-owner in the village:

> I want people to come to the village and my café to relax and forget about stress and their busy lifestyles.
>
> (Interview 14, April 2024)

This sentiment highlights the importance of hospitality and community warmth in shaping the visitor experience. The hospitality sector in the village, which includes cafés, homestays and small businesses, serves as an extension of the community's identity, where guests are not merely passive observers but active participants in village life.

In addition, the incorporation of coffee as a cultural and sensory experience exemplifies the village's dedication to mindful tourism. From cultivating coffee plants to the meticulous roasting and brewing processes, coffee serves as a symbol of patience, craftsmanship and harmony with nature. It embodies the village's philosophy of slowing down and appreciating the journey as much as the outcome, thereby reinforcing the significance of sustainable, small-scale tourism that prioritises depth over volume.

Visitors echo these sentiments, expressing a sense of deep appreciation for the village's serene atmosphere and the warmth of its people. One visitor reflects:

> Everything here is one of my unforgettable memories. I like [the village's] beauty, good atmosphere, good views, delicious local foods, and lovely people.
>
> (Interview 22, April 2024)

These aspirations of locals and visitors highlight the emotional and experiential impact of the village, which extends beyond superficial attractions to leave lasting impressions of peace, fulfilment and cultural enrichment.

Looking ahead, Hmong Doi Pui could reclaim its status as a role model for CBT villages in Chiang Mai region by providing cultural and experiential experiences rather than superficial souvenirs and attractions. Instead of merely serving as a commercial attraction, the village could foster an environment where guests can slow down, reflect and engage deeply with local traditions. This intentional approach to tourism would benefit both visitors and the local community, allowing residents to take pride in their heritage while maintaining control over their own destiny's development.

Ultimately, the preferable future of Hmong Doi Pui as a CBT village is one of balance, authenticity and mutual enrichment. It would be a place where visitors do not merely pass through, but instead, where they pause, connect and carry away with them the wisdom of a slower, more intentional way of living. In doing so, the village would remain not just a destination, but a meaningful experience – one that lingers in the hearts and minds of all who step into its tranquil embrace.

Pathways to Sustainable Tourism in Hmong Doi Pui Village: Strategies and Implementable Interventions

The preferable future of Hmong Doi Pui village relies on carefully designed strategic interventions that align with the cultural, economic and environmental values of the community while addressing key challenges in the current tourism model. Achieving this vision requires balancing the needs of visitors seeking meaningful experiences with the aspirations of the local community to preserve their heritage and maintain control over tourism development. By enhancing immersive cultural experiences, weaving the village's rich coffee heritage into its tourism offerings, and promoting collaboration among stakeholders, Hmong Doi Pui could develop into a leading CBT destination focused on cultural exchange, economic resilience and sustainable growth.

One of the primary challenges currently affecting tourism in Hmong Doi Pui is the perception of excessive commoditisation (Cohen, 1988), where the proliferation of souvenir shops and the prevalence of brief, highly structured visits limit opportunities for authentic cultural engagement. Many visitors express disappointment at the lack of meaningful interaction with the local community and a sense that their experience is superficial, with little to differentiate the village from other CBT destinations in Chiang Mai. This issue is further exacerbated by the fact that most visitors only spend a short amount of time in the village, often no more than an hour, before continuing on to other attractions in the region. In order to transform Hmong Doi Pui into a place where visitors stay longer and engage more deeply with the local culture, a strategic shift towards experience-based tourism is necessary.

Building upon the foundation established by the D-Hope Project, implementing diverse, high-quality workshops can serve as a key strategy for fostering cultural exchange while enhancing the economic sustainability of tourism in the village.

These workshops, designed to align with visitor interests and community expertise, should emphasise traditional Hmong crafts such as textile weaving, embroidery and hemp jewellery making, alongside culinary experiences focused on traditional cooking and herbal medicine preparation. Additionally, workshops centred on agricultural practices, including sustainable farming methods and the integration of coffee cultivation into Hmong livelihoods, would further diversify the offerings available to visitors. Expanding these workshops would ensure that visitors are not merely passive observers of Hmong culture but active participants in a process that allows for deeper understanding and appreciation.

To ensure that these workshops are accessible and integrated into the broader tourism ecosystem, collaboration with tour operators and travel agencies is essential. Rather than relying on visitors to discover these experiences upon arrival, workshops should be available for advance booking through established travel networks. This would allow visitors to plan their participation in workshops before their arrival, increasing the likelihood of engagement and extending the duration of their stay in the village. Tour operators should play a critical role in promoting these experiences by incorporating them into their itineraries, offering curated packages that highlight Hmong Doi Pui as a destination for cultural immersion and slow living.

A possible barrier to engagement in workshops is the language gap between visitors and local community members, which can sometimes deter tourists from participating in interactive experiences. While many locals are eager to share their knowledge and skills, communication challenges can limit the depth of cultural exchange. To address this issue, a system of trained local guides and translators should be implemented to facilitate meaningful interactions between visitors and hosts. These guides should possess not only language skills but also an in-depth understanding of Hmong history, traditions and values, ensuring that they would effectively convey cultural knowledge while respecting the perspectives and experiences of the local community.

Beyond the expansion of cultural workshops, another key intervention for the sustainable development of Hmong Doi Pui would be the deeper integration of its coffee heritage into the visitor experience. Coffee cultivation has played a transformative role in the economic and social evolution of the village, marking a shift from opium farming to sustainable agricultural practices. However, despite the significance of coffee in the daily lives of the Hmong people, its role in tourism remains largely underdeveloped. Many visitors consume coffee in the village's cafés without fully understanding its cultural and historical importance, missing an opportunity for a richer and more meaningful experience. By positioning Hmong Doi Pui as a coffee destination, the village would further differentiate itself from other CBT destinations while providing an additional avenue for economic empowerment.

A multi-faceted approach would be required to develop the village's coffee culture into a compelling tourism experience. The implementation of coffee workshops and guided coffee tours would allow visitors to gain first-hand insight into the process of coffee cultivation, from harvesting and drying to roasting and brewing. These experiences should highlight the unique characteristics of Hmong coffee, including the traditional processing techniques that distinguish it from other regional coffee varieties. Visitors could be invited to participate in hands-on activities, such as picking coffee cherries during harvest season, manually roasting beans or learning about the flavour profiles of different roasting levels through tasting sessions. These interactive elements would enhance visitor appreciation of the coffee-making process while reinforcing the values of patience, craftsmanship and sustainability that define the slow life philosophy.

In addition to structured coffee experiences, a more immersive approach could be taken by developing the concept of a 'Coffee Stay,' where visitors spend an extended period in the village living with coffee farmers and participating in their daily routines. Inspired by successful initiatives in coffee-producing regions such as Colombia and Ethiopia, this concept would offer guests an opportunity to engage with every stage of coffee production while experiencing village life first hand. Such an initiative would attract coffee enthusiasts, eco-tourists and travellers seeking authentic rural experiences, expanding the village's appeal to niche tourism markets.

For visitors with limited time, a more concise version of the coffee experience could be offered through educational coffee tours and mindfulness-based activities such as coffee meditation. By incorporating elements of sensory awareness and slow appreciation into coffee rituals, visitors could engage in a form of mindfulness that aligns with the village's broader philosophy of slow living. These experiences would position Hmong Doi Pui as a unique destination where coffee consumption was not merely an everyday activity but a meaningful and reflective practice that fosters a deeper connection to both nature and community.

Marketing and promotion would play a crucial role in establishing the village's reputation as a coffee destination. A targeted approach should be taken to attract specific visitor segments, including coffee connoisseurs, eco-tourists and professionals in the food and beverage industry. Establishing partnerships with international coffee organisations, barista academies and hospitality schools can further enhance the credibility and visibility of Hmong Doi Pui's coffee experiences. Additionally, digital marketing strategies, including the creation of a 'Coffee Map' that guides visitors through the village's cafés and coffee-related experiences, could enhance accessibility and engagement.

The successful implementation of these interventions would require the active participation of multiple stakeholders, each playing a critical role in ensuring long-term sustainability. The local community must take the lead in workshop facilitation, coffee production and hospitality services, ensuring that tourism remains a community-driven enterprise rather than an externally imposed model. Government agencies such as DASTA should provide infrastructure support, funding opportunities and policy frameworks that prioritise sustainable, small-scale tourism over large-scale commercialisation. NGOs specialising in CBT development could serve as facilitators, offering capacity-building programmes, ethical tourism training and financial literacy workshops to empower community members. Tour operators and travel agencies should act as intermediaries, connecting visitors with the village's cultural and coffee-related experiences while ensuring that tourism flows align with community needs and expectations.

By implementing these strategic recommendations, Hmong Doi Pui could achieve its vision of becoming a leading example of sustainable CBT in Chiang Mai. Expanding and differentiating cultural workshops, integrating coffee into the tourism experience and fostering collaboration between stakeholders would ensure that the village remains an attractive and meaningful destination for visitors while preserving the integrity of its cultural and environmental heritage. Through this approach, tourism would not only serve as a means of economic livelihood but also as a tool for cultural pride, intergenerational knowledge transfer and cross-cultural exchange, allowing Hmong Doi Pui to thrive as a resilient and self-sustaining CBT village in the years to come.

Conclusion

The future of Hmong Doi Pui hinges on its capacity to overcome the challenges presented by commercialisation, limitations in visitor engagement and the need for economic sustainability, while carefully utilising its rich cultural heritage, natural beauty and deeply-rooted traditions. The village faces pressures from rising competition but also has the opportunity to reshape its tourism narrative in line with the principles of slow living, cultural immersion and community-driven development. Through strategic interventions that emphasise meaningful experiences, cultural exchange and economic inclusivity, Hmong Doi Pui has the potential to claim its status as a premier model of CBT for sustainable and responsible tourism in Chiang Mai.

Beyond visitor engagement challenges, economic disparities within the community persist, with many tourism-related benefits concentrated in retail businesses rather than diversified across multiple sectors such as hospitality, agriculture and experiential tourism. The full potential of CBT has not yet been realised, as many locals remain in passive roles rather than actively shaping tourism offerings. Without deliberate planning, there is a risk that Hmong Doi Pui could become excessively commodified, sacrificing the very authenticity that visitors and residents desire. Additionally, as competition with other CBT villages and emerging tourism destinations grows, differentiation becomes critical to ensuring the village's long-term appeal.

Despite these challenges, Hmong Doi Pui possesses significant opportunities to transition towards a more sustainable and culturally enriching tourism model. Its strong sense of community identity, deep-rooted traditions and connection to nature offer a robust basis for a tourism experience that prioritises quality over quantity. By embracing slow living and immersive cultural exchange, the village can transcend mass tourism dynamics and establish a tourism model that respects both the cultural integrity of the community and the expectations of responsible tourism travellers.

The proposed strategic interventions – Crafting Culture and Coffee Calms – would serve as key mechanisms for this transformation, ensuring that tourism becomes a tool for economic empowerment, cultural preservation and meaningful visitor engagement. Crafting Culture addresses the lack of structured, interactive cultural experiences by expanding and formalising hands-on workshops in traditional crafts, culinary heritage and sustainable agriculture. These workshops would provide an alternative to passive tourism, allowing visitors to actively engage with local artisans, farmers and knowledge-holders. They would also ensure that tourism revenue was more equitably distributed, offering direct economic benefits to those involved in cultural production and heritage transmission.

Coffee Calms, in turn, positions Hmong Doi Pui as a premier local coffee production destination in Thailand, leveraging its rich coffee-growing heritage alongside the principles of mindfulness and sustainability to craft an experience-driven tourism product. Through guided coffee experiences, immersive homestays with coffee farmers and educational coffee tours, visitors would gain a deeper understanding of the cultural and environmental significance of coffee in the village. By integrating coffee culture into the broader slow-living philosophy, this intervention would further reinforce the village's identity as a place of reflection, relaxation and mindful connection with nature.

To successfully implement these interventions, collaborative stakeholder engagement is essential. The local community should take ownership of tourism development, ensuring that economic benefits are shared equitably and that cultural traditions are protected from over-commercialisation. Tour operators and travel agencies play a critical role in promoting workshops and coffee experiences, integrating them into curated travel itineraries that encourage longer stays and deeper cultural engagement. Government agencies, such as DASTA, must provide infrastructure support and financial incentives, ensuring that the village has the necessary resources to develop high-quality, community-led tourism initiatives. NGOs and academic institutions can serve as facilitators, offering expertise in capacity-building, ethical tourism and cross-cultural communication. Through these partnerships, Hmong Doi Pui can ensure that its tourism development trajectory remains aligned with principles of sustainability and community empowerment.

The preferable future of Hmong Doi Pui is one in which tourism enhances rather than undermines community identity, fostering a harmonious balance between economic opportunity, cultural preservation and environmental stewardship. By shifting away from passive, commercialised tourism towards experiential and interactive models, the village could offer visitors an alternative to mass tourism, providing meaningful, slow-paced and immersive experiences that foster genuine connections between travellers and the local community. This vision ensures that the cultural and natural assets of Hmong Doi Pui remain intact for future generations, securing long-term sustainability in both tourism and community well-being.

References

Betgem, F. (2022). Doi Pui Hmong Village – Green trails. Available at www.green-trails.com/doi-pui-hmong-village/ (Accessed 26 February 2025).

Boonratana, R. (2010). Community-based tourism in Thailand: The need and justification for an operational definition. *The Kasetsart Journal: Social Science*, *31*(2), 280–289.

Chiang Mai Guideline. (2018). Doi Pui Hmong Hilltribe Village. Available at https://chiangmaiguideline.com/doi-pui-hmong-hilltribe-village/ (Accessed 26 February 2025).

Cohen, E. (1988). Authenticity and commoditization in tourism. *Annals of Tourism Research*, *15*(3), 371–386.

Ditta-Apichai, M., Sroypetch, S., & Caldicott, R. W. (2024). A critique of community-based tourism development: The comparative case of Betong and Pho Tak Districts, Thailand. *Community Development*, *55*(1), 67–84.

Han, G., Wu, P., Huang, Y., & Yang, Z. (2014). Tourism development and the disempowerment of host residents: Types and formative mechanisms. *Tourism Geographies*, *16*(5), 717–740.

Scheyvens, R. (2002). *Tourism for Development: Empowering Communities*. England, Pearson Education.

United Nations Development Programme (UNDP). (2022). Community-based tourism: Empowering local champions for sustainable tourism in Thailand. Available at www.undp.org/thailand/blog/community-based-tourism-empowering-local-champions-sustainable-tourism-thailand (Accessed 26 February 2025).

Zhu, L., Zhan, L., & Li, S. (2021). Is sustainable development reasonable for tourism destinations? An empirical study of the relationship between environmental competitiveness and tourism growth. *Sustainable Development*, *29*(1), 66–78.

Part IV

11 Conclusions

We live in a 'VUCA' world, which stands for volatile, unpredictable, complex and ambiguous (Lubowiecki-Vikuk & Sousa, 2021). This means we must accept the reality of constant, unpredictable change and a lack of understanding about how interrelated components shape and evolve, resulting in an environment that is difficult to forecast, interpret and analyse (Major & Clarke, 2022).

Scenario planning is a prominent well-used strategy for dealing with uncertainty in strategic decision-making (Postma, Hartman & Yeoman, 2024, p. 10). Scenario planning, which began as an open and uncertain attempt, has grown in many schools and is now widely used in research and practical circumstances (López-Rodríguez et al., 2023). Scenario planning is referred to by various titles in the literature, including scenario development, scenario planning and scenario thinking (Oner, 2010; Varum & Melo, 2010), as well as participatory scenario planning (Poskitt, Waylen & Ainslie, 2021). The future offers unlimited opportunities for sustainable tourism research, providing rigorous and diverse quantitative and qualitative methodologies which can be devised to test and compare findings. Since the future is constantly evolving, ongoing research will undoubtedly reveal new market trends and industry changes. The focus of this book has been designed to illustrate three different ways of thinking about the future, using Henchey's typology of futures (1978): (1) *possible future* (what may be) encompasses everything that can be imagined, no matter how unlikely; (2) *probable future* (what will likely be) is what will likely happen and is based on an examination of the current situation and an appraisal of likely trends and future developments; and (3) *preferable future* (what should be) is what is wanted to happen and falls in the realm of organisational or societal visions that "move reality beyond the present toward the best that can be" (Hancock & Bezold, 1994, p. 25).

Approach and Methods

As noted earlier, three different locations were selected for study, parts of Melbourne, Australia; Penang, Malaysia; and Chiang Mai in Thailand. These research sites represent destinations in different stages of development and at different scales

DOI: 10.4324/9781003570240-15

of operation. In all three locations the approaches and methods used were similar, involving data collection (both historical and current), interviews with (mostly academic) experts and with local stakeholders, observations from field work and netnography to obtain wider external viewpoints and opinions. Coding of data led to the identification of specific themes, opportunities and challenges and provide the information on which the conclusions are based.

Findings

This edited collection addresses a research approach for understanding the future of destinations.

Possible Future

The collected conclusions in Part I of this book provide a comprehensive examination of several areas in Melbourne, Australia, along with the ramifications for selected stakeholders. Chapter 2 focuses on the urban developments in Southbank in the context of the possible futures of micro- and small-sized enterprises (MSEs) in this area. Southbank has been subdivided into seven precincts to clarify the differences and similarities relating to the developments and impacts between the areas. The research involved examining businesses operating in branches that are either directly or indirectly related to tourism. The urban development projects of the future emphasise the expansion of art and greenery into the surrounding precincts of Boyd, Malthouse and South Melbourne. This is expected to give these areas more character and a fresh appeal and offer new opportunities for MSEs. In addition to becoming a more attractive area for residents, the developments will likely also attract more tourists. Consequently, it is expected that tourism-related businesses in the now touristic area will expand to the surrounding precincts behind the riverfront to take advantage of the new markets.

The third chapter focuses on the dynamics of and developments in Melbourne's Central Business District (CBD) that may have an impact on the future of MSEs there. The researchers examined the CBD environment using an interdisciplinary method, employing several paradigms and considering numerous variables pertinent to the field. Possible future developments that are applicable in the setting of specific dynamics found in the researched area of core CBD are presented. The findings show that population expansion and a rise in the number of tourists in the central CBD will create opportunities for MSEs to expand as a business. The challenges that will be faced by MSEs and the CBD area include higher rents which make accessing an affordable site difficult, growing social issues, friction between tourists and residents because of the uptake of Airbnb rentals, constructions causing disruptions and limiting visibility and access to MSEs behind the scaffolding on a short-term basis. However, these developments also provide new opportunities and challenges in the future. Competition between medium and large organisations (MLOs) and MSEs will arise due to the construction of new transport stations, however, new technologies will enhance accessibility, providing opportunities for experiencing the city and further opportunities for MSE development in the core CBD.

Chapter 4 reveals the possible futures of Docklands, in this case, focusing on the strategic position of MLOs as well as residents. Docklands is a relatively young precinct, developed from 1990 onwards. It is surrounded by the Yarra River and located west of the CBD. The Marvel Stadium and the Victoria Harbour are located in the centre of the precinct, large organisations are situated in the south, while the north has the entertainment area. Docklands, with a high proportion of foreign nationalities, is multicultural.

The connectivity of Docklands is difficult to explain, as Docklands is surrounded by the Yarra River, the surface railway and major roads, and is divided by the Yarra River, which hinders commuters from travelling easily from different suburbs into Docklands. Many residents and workers have begun using bicycles as a mode of transport, and possible future access options include bike paths separated from pedestrian areas, commuter ferries, a metro tunnel or a tram bridge across the Yarra River. Such improvements would help the many employees of large organisations as well as residents with commuting and connecting with other precincts. The Marvel Stadium as a dominant structure, in the centre of Docklands, complicates the connection between Docklands and the CBD because it enables many large events to be staged. Hotels are relatively dependent on these events, as visitors stay overnight. The future of this stadium is uncertain as it might be remodelled, relocated (within the Docklands or out of it) or stay the same. If it changes location within the Docklands it could reduce the connectivity problem with the CBD, making the Docklands a more attractive place to live and work. As a result of COVID-19 many of the workers in the office buildings in the south of the area now work from home, and with fewer people commuting to and from Docklands, employees working from home might become the new normal.

Future urban development projects in Melbourne potentially will enhance the character and appeal of Melbourne as a tourism destination and will provide new opportunities and attractions. These developments are likely to draw more tourists as well as make the neighbourhoods more appealing to residents, and as a result, it is envisaged that tourism-related firms in the newly touristic area will spread into the neighbouring precincts. The anticipated increase in the number of tourists in Melbourne should generate chances for numerous businesses to expand. In the future, similar issues and patterns will emerge, such as population growth, higher rents that make it difficult for residents and businesses to find an affordable site, rising social issues, friction between tourists and residents as a result of the popularity of Airbnb rentals and construction disruptions that limit visibility and access to businesses. New technology will boost accessibility, allowing people to explore the city and visit new developments in Melbourne.

Probable Future

Part II shifts to the probable future for destinations. What is 'most likely' to happen at the destination (Penang, Malaysia) and what will be the consequences for stakeholders? In this section, we identify certainties and uncertainties at the destination level and identify the global drivers of change (changes in demographics, economic activity, consumer preferences and technology). Chapter 5 examines the probable future of George Town in Penang and outlines one probable future scenario for tourism development, by exploring the current dynamics of tourism and stakeholder perspectives. The future scenario drawn in this chapter includes expected impacts and resulting consequences for both MLOs as well as public organisations and the responses that they might have. The findings reveal that a division and lack of cohesion among stakeholders regarding preferences and strategies are evident, with no foreseeable resolution in sight. This gap persists because of the inherent conflict between promoting quality of life and accommodating tourism growth in George Town. As a result, the probable future of the city appears to be one where tourism expansion is likely to come at the expense of quality of life of residents and the natural environment. Moreover, the diverse array of tourists visiting the area will have varied impacts on stakeholders, leading to differing focuses and priorities. Some stakeholders may prioritise managing overcrowding, or target specific tourist markets, while others will persist in tackling climate change issues and creating a liveable area.

Chapter 6 focuses on Gurney Drive (divided into 'local,' 'medical' and 'Gurney Bay' areas), which is the most rapidly developing part of the Malaysian island of Penang. Its reclaimed land is expected to have a bright future. Currently, Gurney Drive offers new luxury hotels, large malls and advanced hospitals for its visitors, and the focus of the research is on MLOs that influence and are influenced by these developments at Gurney Drive. The stakeholder groups are 'development and construction,' 'community and living' and 'tourism and hospitality.' In the next five years, Gurney Bay is set for significant transformation, becoming a hub for leisure activities and community gatherings. Despite optimism over positive changes and increased local satisfaction, challenges such as waste management and traffic congestion are likely to persist. Luxury tourism is likely to continue to thrive and develop further to attract high-spending tourists, while developments in specialised medical care and treatments will strengthen Penang's desired position as a leading destination for medical tourism. Gurney Bay's leisure appeal will increasingly target wealthier individuals and international investors, supported by expansions in infrastructure and hospitality services, boosting the state's and country's economy. Hotels will experience increased competition in the luxury hospitality sector and an overall rise in leisure tourists, and hospitals will record a rise in medical tourists and continuously grow their facilities further. Over the next five years, construction companies will be involved in the development of new retail facilities, hotel expansions and the continuation of the Andaman Island development project, while real estate companies will experience higher demands for housing and apartments along Gurney.

Chapter 7 examines Penang Hill, a popular tourist destination offering diverse attractions and the focus here is on implications for MLOs. Findings reveal that features, assets and potential do match the wishes of the current market, although the funicular train cannot handle the number of visitors that want to visit the attraction. In the future, a cable car will be completed that should resolve this issue to some extent. Managing the visitors at the top of the hill will become more of a challenge because of its accessibility and the closeness of the hill to George Town, meaning it is going to attract more visitors. This also means more people will visit the Biosphere Reserve (Habitat) atop the hill which is focused on the preservation of, and education about, the natural world. Habitat's aims differ from those of other businesses on Penang Hill, requiring a delicate balance between conservation and economic development. In the next five years, these businesses need to work together efficiently to ensure a balance between their priorities to avoid an imbalance or conflict.

Preferable Future

Part III moves to the preferable future and focuses on Chiang Mai, Thailand, where we discuss the numerous issues Chiang Mai faces. The aim is to propose recommendations to the various players to assist them in attaining the destination's preferred future. The current strategic positions of the stakeholder groups are identified and related to the destination area, its features, assets, potential and current markets. This analysis allows the definition of the most preferable development of the area over the next five years.

Chapter 8 examines Chiang Mai's Old Town, a historical and cultural landmark in Northern Thailand, which stands at a critical juncture in its development as a tourism hub. While tourism has played a central role in shaping the city's economy, rapid growth, globalisation and urbanisation pose challenges to its cultural heritage, environmental sustainability and the well-being of local communities. The preferable future of Old Town, Chiang Mai, is determined by analysing historical and current developments, key stakeholders and emerging challenges. The preferable future is created based on stakeholder aspirations, with specific attention given to tourism development, sustainability and cultural preservation. The chapter also offers interventions and recommendations for policymakers, businesses and local communities in order to balance economic growth with heritage conservation. By implementing strategic initiatives such as community engagement, responsible tourism practices and technological innovation, Old Town, Chiang Mai, can establish itself as a model for sustainable tourism while retaining its unique character identity.

Chapter 9 analyses the evolving landscape of Nimman, Chiang Mai, and its trajectory towards a preferable future amid rapid urban transformation. Once a quiet residential neighbourhood named after the Nimmanhaemin family – who contributed land for the establishment of the Chiang Mai University – Nimman has undergone a profound shift, emerging as a dynamic urban hub characterised by international cuisine, co-working spaces and a burgeoning medical tourism sector. Its cosmopolitan appeal has positioned it as a key destination for Chinese tourists, digital nomads and investors, reinforcing its status as Chiang Mai's modern commercial epicentre. By analysing both public and private assets, the study evaluates how strategic interventions may influence the district's long-term viability. Findings suggest that Nimman will continue to attract Asian tourists, digital nomads and expatriates, drawn by its international atmosphere and commercial vibrancy. However, its future remains subject to political and environmental uncertainties, which may influence its long-term resilience. Despite persistent political and economic volatility, Nimman's development trajectory suggests an increasing trend towards commercialisation at the expense of residential use, further solidifying its identity as Chiang Mai's leading modern district. By critically examining the forces driving this transformation and their socio-economic implications, this chapter contributes to broader debates on tourism-driven urban change and the challenges of balancing destination development with local sustainability in Thailand.

Chapter 10 investigates the preferable future for Hmong Doi Pui village and outlines strategic pathways to achieve this. The findings highlight a unified vision to transform the village into a community-based tourism (CBT) destination focused on slow living, strongly anchored in Hmong cultural heritage, with an emphasis on coffee production, nature and traditional crafts. This vision seeks to attract visitors who are interested in immersive and meaningful experiences that foster engagement with the local community and its natural surroundings. To realise this vision, two primary strategic interventions are proposed. 'Crafting Culture' introduces workshops led by artisans, facilitating cultural exchange and the preservation of traditional craftsmanship. 'Coffee Calms' aligns with the village's philosophy of slow living, providing coffee-centric experiences that include guided stays, workshops and meditation sessions. These initiatives are designed to enrich visitor experiences, redefine and retain the village's identity and promote sustainable tourism. The chapter emphasises the vital importance of cultural preservation and community involvement in creating a sustainable and appealing future for Hmong Doi Pui, benefiting both residents and visitors.

While tourism has played an important role in defining the city's economy, fast growth, globalisation and urbanisation pose threats to its cultural legacy, environmental sustainability and the well-being of local residents. Chiang Mai can develop itself as a paradigm for sustainable tourism while keeping its distinct character. Chiang Mai highlights the critical role of cultural preservation and community interaction in ensuring a sustainable and desirable future for both residents and visitors. Chiang Mai will continue to attract Asian tourists, digital nomads and expatriates due to its cosmopolitan atmosphere and commercial activity. However, its future remains uncertain due to political and environmental factors, which may have an impact on its long-term resilience. The future will require balancing local perspectives while addressing present and potential challenges.

Commonalities and Conclusions

While the three locations are very different in many respects certain commonalities relating to their futures emerge from the research. One factor is scale. Melbourne is a large multicultural international city, where the numbers of those engaged in all facets of leisure (tourism, recreation and leisure pursuits) run into the hundreds of thousands on a daily basis, interacting and sharing the many leisure resources and opportunities. The likely future is influenced heavily by local (Melbourne) and state (Victoria) policies and developments, some driven by external agencies and forces and also by major features in the leisure scene which affect not only participation but also mobility and access. Almost no single facility is capable of radically changing the nature or extent of development because of the size of the city and long-established infrastructure and planning systems. The other two locations, Penang and Chiang Mai are much smaller, more recently developed in terms of tourism and leisure use, and have higher degrees of uncertainty about their futures, partly because being smaller in terms of investment and visitation, single features or occurrences can still have very significant impacts on future patterns of development. External policies relating to the growth of medical tourism in Penang is one example which could make sizeable differences in parts of the city and the nature of future development. In Chiang Mai there is less coherence in terms of clear overall senior government policy direction, the result of which can be seen in the less organised and more informal development of opportunities and lack of focus over development themes.

The second common major feature is the importance of stage of development. As noted, Melbourne is a long established city with a century or more of tourist visitation and associated development, along with the growth of a large number of leisure opportunities, thus to some extent, predetermining the future for MLOs and MSEs alike. While Penang has a tourism history also a century or more in age, for much of that time the tourism was by the colonial elite, and independence then altered the nature and level of activity, and it is only relatively recently that tourism, particularly international tourism, has developed to a significant level. Thus, there is more scope and opportunity for individual enterprises to enter the market and even change the focus of tourism, as the medical element has done. Chiang Mai is much more recently involved in international tourism and clear foci have not yet really emerged in any planned way. This leaves a great deal of opportunity for enterprises at all scales to enter the market and for that market to be broadened into new areas and new themes created, such as that proposed related to coffee production. In that sense, the smaller and more recent the involvement in tourism, the greater the potential range of opportunities for identification and development of new and preferred directions. However, such a situation is also vulnerable to influence from large and powerful exogenous forces, from other countries, for example China, from other parts of its own country and from existing large enterprises.

All futures, of course, depend heavily on stability of current trends and practices. The advent of COVID-19 demonstrated the impacts of a sudden catastrophic change in circumstances, both in these locations and globally. Other issues such as the impact of climate change, with possible restrictions on long haul air travel or increases in all travel costs, could dramatically change the realism of the conclusions drawn here.

References

Hancock, T., & Bezold, C. (1994). Possible futures, preferable futures. *Healthcare Forum Journal*, *37*(2), 23–29.

Henchey, N. (1978). Making sense of future studies. *Alternatives*, *7*(2), 24–27.

López-Rodríguez, M.D., Oteros-Rozas, E., Ruiz-Mallèn, I., March, H., Horeca-Milcu, A.I., Heras, M., Cebrián-Piqueras, M.A., Andrade, R., Lo, V.B., & Piñeiro, C. (2023). Visualizing stakeholder's willingness for collective action in participatory scenario planning. *Ecology and Society*, *28*(2), 5.

Lubowiecki-Vikuk, A., & Sousa, B. (2021). Tourism business in a VUCA world: Marketing and management implications. *Journal of Environmental Management and Tourism*, *12*(4), 867–876.

Major, J., & Clarke, D. (2022). Regenerative tourism in Aotearoa New Zealand – A new paradigm for the VUCA world. *Journal of Tourism Futures*, *8*(2), 194–199.

Oner, A. M. (2010). On theory building in foresight and future studies: A discussion note. *Futures*, *42*(October), 12. Available at www.sciencedirect.com.helicon.vuw.ac.nz/science/journal/00163287/42/9 (Accessed 10 December 2024).

Poskitt, S., Waylen, K.A., & Ainslie, A. (2021). Applying pedagogical theories to understand learning in participatory scenario planning. *Futures*, *128*, 102710.

Postma, A., Hartman, S., & Yeoman, I. (2024). Why the future is important in an uncertain world: The role of scenario planning to create the future of tourism. In A. Postma, S. Hartman, & I. Yeoman (Eds.), *Scenario Planning and Tourism Futures: Theory Building, Methodologies and Case Studies* (pp. 1–7). Bristol: Channel View.

Varum, C. A., & Melo, C. (2010). Directions in scenario planning literature – A review of the past decades. *Futures*, *42*(4), 355–369.

Index

Note: Page numbers in *italics* indicate figures and in **bold** indicate tables on the corresponding pages.